SpringerBriefs in Materials

For further volumes:
http://www.springer.com/series/10111

Jameel Ahmed · Mohammed Yakoob Siyal
Freeha Adeel · Ashiq Hussain

Optical Signal Processing by Silicon Photonics

 Springer

Jameel Ahmed
Freeha Adeel
Ashiq Hussain
HITEC University
Taxila
Islamabad
Pakistan

Mohammed Yakoob Siyal
School of Electrical and Electronic
    Engineering
Nanyang Technological University
Singapore
Singapore

ISSN 2192-1091   ISSN 2192-1105   (electronic)
ISBN 978-981-4560-10-8   ISBN 978-981-4560-11-5   (eBook)
DOI 10.1007/978-981-4560-11-5
Springer Singapore Heidelberg New York Dordrecht London

Library of Congress Control Number: 2013944760

Printed on acid-free paper

Springer is part of Springer Science+Business Media (www.springer.com)

Preface

Optical fibre communication is the most promising technology to meet the ever growing need of bandwidth in the present technological era. It not only provides high capacity transmission but also offers flexibility in transportation of data on a network. The major obstacle in further increase in transmission capacity is the use of electronics in signal processing functions and signal amplification. Many researchers and scientists came up with different solutions to eliminate electronics and presented all-optical signal processing techniques. Among these techniques, the use of Self Phase Modulation (SPM), Cross Phase Modulation (XPM), Raman Scattering and Four Wave Mixing (FWM) are well considered but FWM is more attracting due to its inherited benefits. Silicon-on-Insulator (SOI) waveguide devices are emerging to realize any modern optical signal processing scheme. The recent technological improvements in silicon photonics is the main driving force behind the success of these devices. Using nonlinear optical phenomenon in silicon wires and their compatibility with CMOS devices provides the platform for integrated photonic devices. All optical signal processing devices are being investigated and explored at present; however the chip-scale solution provided by silicon photonic is the preferred solution. In this book, the authors have intended to present the summary of their research work in this area. The book focuses on achieving successful optical frequency shifting by Four Wave Mixing (FWM) in silicon-on-insulator (SOI) waveguide by exploiting a nonlinear phenomenon.

This book presents the basic facts, concepts, principles and applications of the nonlinear effects inside the Silicon-on-Insulator waveguide. In this research, in order to achieve optical frequency shifting by FWM in SOI waveguide, a nonlinear phenomenon has been applied successfully. Further, the text is aimed at making graduate students understand the nonlinear effects inside SOI waveguide and possible applications of the latter in this emerging and dominating area of research. The devices manufactured using this technology and the inherent obstacles of the structure faced for some of the fruitful applications are also discussed in this book. The FWM process in SOI waveguides is explained with an emphasis on the effects of two-photon absorption and the consequent free-carrier effects. The optimization of frequency shifting, conversion efficiency and the effect of different parameters on conversion efficiency are also taken into account. Keeping in view the anticipated curiosity of readers, the complete simulation and its subsequent results have explicitly been discussed and demonstrated with illustrating diagrams. It will certainly be more interesting for the readers to know that how the all-optical frequency shifting using single pulsed pump light takes place. The book further encompasses the study of losses due to two photon and free carrier absorption and how to overcome these losses with SOI waveguides.

The book consists of nine chapters in total in which Chap. 1 introduces the subject area along with its significance in the diversified applications.

Chapter 2 describes the research-based past and future trends of the optics. Various research aspects in the area of silica-glass fibre have substantially been explored and pointed out for future trends. This chapter incorporates to a significant extent the current and updated research to offer the foundation for the work that has been done in this field.

Chapter 3 discusses the standards of optical communication and network theory along with their applications and significance in the respective areas. It also examines the characteristics of optical communication systems by discussing different communication media. The devices being used and considered for future consideration are also discussed and summarized for optical networks. Chapter 4 focuses on the gradual and time-needed advancements made in the field of photonics. It also highlights the evolution taken place in the devices which are being used and are being prepared for future trends as well. The traditional and non-traditional motivations regarding selection of silicon as the enabling material for the usefulness of photonics and SOI waveguides are also presented with great emphasis and depth.

Chapter 5 highlights the effects of nonlinearity in optical fibre communication link. Desirable effects of nonlinearity in general and undesirable effects in particular in an optical fibre are enumerated with examples. Cross Phase Modulation (XPM) and high nonlinearity glasses along with their advantages and disadvantages are spelled out with the help of their theories and application in practical systems.

Chapter 6 describes how supercontinuum (SC) sources are replacement of white light sources. Supercontinuum generation in optical fibre, pumped by different sources, which include pumping with femto second (fs), picoseconds (ps) pulse sources and continuous sources are reviewed in this chapter. The nonlinear Schrödinger equation is used to discuss the spectral broadening or SC generation. The chapter also shows that SC generation has been proved very effective and some of its applications are very promising for future ultra-high bandwidth networks.

Chapter 7 provides a theoretical model for pulse propagation inside an SOI waveguide. The Four Wave Mixing (FWM) process in SOI waveguides is discussed with an emphasis on the effects of two-photon absorption and the consequent free-carrier effects. All optical wavelength conversion and optical signal processing along with requisite devices is illustrated with examples exclusively.

In Chap. 8, Four Wave Mixing (FWM) and its types are elaborated in detail. It also describes the mathematical equations which provide basis for mathematical modelling and subsequent realization on various platforms. In order to achieve optical frequency shifting by FWM in silicon-on-insulator (SOI) waveguide, a nonlinear phenomenon has been presented. The FWM process in SOI waveguides is also discussed with an emphasis on the effects of two-photon absorption and the consequent free-carrier effects.

In Chap. 9, in order to achieve optical frequency shifting by FWM in SOI waveguide, a nonlinear phenomenon has been applied successfully by exploiting the Simulink MATLAB® tool. The FWM process in SOI waveguides is the main focus of the work with an emphasis on the effects of two-photon absorption and the consequent free-carrier effects. The optimization of frequency shifting, conversion efficiency and the effects of different parameters on conversion efficiency are also taken into account throughout the time-consuming simulation. The simulated results show the successful all-optical frequency shifting by using signal pulsed pump light and also deal with the losses due to photon absorption and free carrier absorption.The simulated results conclude what numerical values of the different parameters lead to optimal results and how to decrease the undesired losses with SOI waveguide.

In addition to strong understanding of Matlab®, the reader is expected to have prior knowledge of signal processing functions, signal amplifications and familiarization with Self-Phase Modulation, Cross Phase Modulation, Raman Scattering and Four Wave Mixing, etc.

We are highly indebted to our colleagues who extended their cooperation wherever and whenever we looked around, particularly Ms. Menna Nawaz. We appreciate and acknowledge the positive response and forward-looking approach of Dr Ramesh Nath Premnath, Springer Asia throughout our interactions with him. We are reasonably confident in presenting this work, which is the outcome of graduate level research, and hope that it will be a reasonable contribution for the readers in the subject area. However, constructive criticism and suggestions for improvement will warmly be welcomed.

Jameel Ahmed
Mohammed Yakoob Siyal
Freeha Adeel
Ashiq Hussain

Contents

Chapter 1
Introduction

The ability and desire for constant communication between human being is the distinguishing characteristic of mankind. Human beings have always been interested in finding ways to communicate among each other through numerous ways. Their ability to use language to build words, combine these into significant sequences and then articulate them through speech that makes them the most powerful communicators on the planet has always been increasing with the passage of time. Communication between humans can be divided into two different categories by analyzing the evolution and the consequences of communication through history [1]. The first category is the utilitarian communication; it has enabled the rise of the human civilization as a whole through the exchange of concepts, ideas, and knowledge over time. On the other hand, this category has also been used successfully throughout history as an important tool in achieving and maintaining the supremacy and advantages of certain civilizations relative to others. That is why the communication technologies have always been of strategic importance. They enable the human beings to prove themselves more civilized and cultured as compared to their predecessors. The second category of communication is the leisurely communication; it fulfills the need for humans to be heard and understood, to start and maintain the relationships with other humans and also to entertain themselves. Thus, analyzing and categorizing the evolution, human beings first distinguish themselves from others by learning new things and then they want to transfer the knowledge they got, to other people around. This stems from the essence of the human nature.

Communication or more specifically the electrical or optical communication may be defined in broader way as transferring the information between any two points. While transferring this information to distant points usually needs proper communication system(s). In these communication systems the information is generally transferred by modulating or superimposing the modulating signal on an electromagnetic wave. Subsequently, the information signal becomes modulated carrier and then this carrier is transmitted to desired destination where the original signal is retrieved by demodulation process. The transmission is carried on some

J. Ahmed et al., *Optical Signal Processing by Silicon Photonics*, SpringerBriefs in Materials, DOI: 10.1007/978-981-4560-11-5_1, © The Author(s) 2013

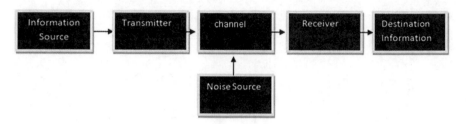

Fig. 1.1 Basic block diagram of communication system

transmission medium, which is used to carry the information from transmitter to receiver (Fig. 1.1).

The visible optical carrier waves or light for communication has been in use frequently over the decades. Simple systems which include signal fires, reflecting mirrors and, signaling lamps used recently are providing very successful, if restricted, information transfer. An optical fiber communication system is the basic form of any type of communication system. In electrical communications the electrical signal provided by the information source, is usually resulted from a non-electrical message signal, to a transmitter comprising electrical and electronic components which are capable of converting the signal into an appropriate form for propagation over the transmission medium. This is often achieved when a carrier, which may be an electromagnetic wave, is modulated. The transmission medium however can comprise a pair of wires, a coaxial cable or a radio link through free space, which is used to transmit the signal to the receiver. It is receiver, where the signal is transformed into the original demodulated electrical information signal before passing to the destination point. However, the obvious thing is that in any transmission medium attenuation is occurred to the signal, or it may suffer losses and degradation occurs due to contamination caused by random signals and noise, also distortions are possibly imposed by different mechanisms within the medium itself. Therefore, in any communication system there is a maximum permitted distance set as threshold between the receiver and the transmitter beyond which the system effectively restrict the intelligible communication. For effective long haul applications these factors demand the installation of repeaters or line amplifiers at intervals, in order to get rid of signal distortion and to boost signal level before transmission process is sustained down the link.

The modulation of an optical carrier may be done using either an analog or digital information signal. In the system that is modulated with analog signal, the variation of the light emitting from the optical source in an uninterrupted manner. The digital modulation, however, involves distinct changes in the light intensity. Analog modulation is although easier and simpler to implement with an optical fiber communication system is not much efficient, it also requires a signal-to-noise ratio at the receiver end far greater than digital modulation. Also, the linearity needed for analog modulation is not always provided by semiconductor optical sources, especially at high modulation frequencies. For these reasons,

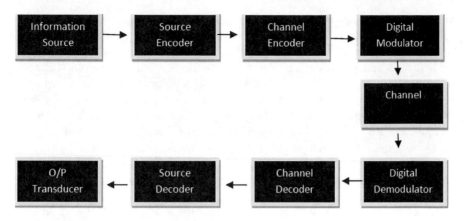

Fig. 1.2 Basic diagram of digital communication

analog optical fiber communication links are generally limited to shorter distances and lower bandwidth operation than digital links (Fig. 1.2).

Communication while using an optical carrier wave guided along a glass fiber has a number of awfully striking features, several of which were apparent when the technique was originally born. Furthermore, the advancements in the technology have surpassed even the most positive predictions; it has created many additional advantages. Hence it is functional to consider the qualities and special features offered by optical fiber communications over more usual mediums for electrical communications. The optical carrier frequency in the range 176 to 375 THz (generally in the infrared region) yields much greater probable transmission bandwidth than conventional metallic cable systems, or even millimeter wave radio systems. Indeed, today the typical bandwidth multiplied by length product for an optical fiber link incorporating fiber amplifiers was 5,000 GHz km in contrast with the typical bandwidth–length product for coaxial cable of around 100 MHz km. Hence at this time optical fiber is already signifying more than a factor of 50,000 bandwidth improvement over coaxial cable while also providing this superior information-carrying capacity over much longer transmission distances.

Integrated technology for optical devices has developed within optical fiber communication to make fabrication of a complete signal possible on a single chip. Integration for such devices has become a confluence of several optical or photonic disciplines. Both Integrated optics (IO) and Integrated Photonics (IP) technologies can be made distinguishable from one another by mean of control of the optical devices. The IO terminology is determined by electronic control of the optical devices and operation of IP devices is controlled by photons. The IP does not consider any optoelectronic conversion hence this technology is also termed as 'all-optical' too. Both IO and IP use planar waveguide technology in order to provide the interconnections between optical components that include the basic components for guiding and control of optical signals. IP technology, however, makes the fabrication of subsystems possible and systems can be considered as a single substrate. Thus integrating both active and passive devices are monolithically onto a single substrate resulting a multilayered integration then these devices are normally referred to as IP devices, while when both active and passive elements fabricated as individual devices are interconnected together they form larger IO devices or circuits. Thus IP can also be seen as a process for the miniaturization and integration of optical systems on a single substrate, and therefore it is considered as a further development of IO technology, not necessarily as an alternative technology. Both IO and IP allow processing to be performed on optical signal by requesting to provide an alternative to the conversion of an optical signal back into the electrical system earlier. Hence thin transparent dielectric layers on planar substrates which act as optical waveguides are used in IO and IP to produce small-scale and miniature optical components and circuits.

The developments in IO have crossed the stage where both signal processing and logic functions can be physically imagined. Furthermore, such devices are considered as the building blocks for future digital optical computers. Nevertheless, quite a number of these devices can be combined and closely linked with developments in light wave communication employing optical fibers.

A most important factor in the development of IO is that it is highly incompatible with multimode fiber systems because it is basically based on single-mode optical fiber. In fact IO did not make a major contribution to early deployed optical fiber systems. The initiation, however, of single-mode transmission technology further motivated work in IO to supply devices and circuits for these more developed third-generation systems. In addition, the continuous expansion of single-mode optical fiber communications has helped in creating a developed market for such IO components. More likely the new generations of optical fiber communication systems employing coherent and possibly soliton transmission will lean heavily on IO and IP techniques for their implementation.

The other major feature provided by optical signals while interacting within a responsive medium is the capability to make use of light waves of distinct frequencies (or wavelengths) within the same guided wave channel or device. Such frequency division multiplexing makes possible the information transfer capacity far better than anything offered by electronics. Moreover, in signal processing language it allows parallel access to information points within an optical system. This option for prevailing parallel signal processing coupled with ultrahigh speed operation offers remarkable potential for applications within both communications and computing.

The devices of interest in IO and IP are often the counterparts of microwave or bulk optical devices. These include junctions and directional couplers, switches and modulators, filters and wavelength multiplexers, lasers and amplifiers, detectors and bistable elements. It is envisaged that developments in this technology will provide the basis for the next generation of optical networks. The technology associated with the design and fabrication of IP circuits and devices highly depends upon different factors that mostly result from the characteristics of the substrate material on which the various devices are to be fabricated. The IP process may require serial, parallel or hybrid integration of independent devices. In serial integration of devices, different elements inside optical chip can be interconnected in an uninterrupted manner and therefore side-emitting, edge-emitting, or conducting optical devices can be eagerly integrated on the same substrate. In the case of parallel integration, the chip is constructed by developing the columns of devices, doing so the surface- or bottom-emitting devices can effectively be used whereas in the case of hybrid integration IP technology the devices are fabricated using both serial and parallel integration on the same substrate. Additional elements can be developed separately or directly attached to the IP circuit to access the control of the optical signals. In addition, both active and passive devices may be required to be located on the same substrate and therefore hybrid IP integration demands multilayered IP circuits and components to be produced on a single substrate such that they must be compatible with three-dimensional structures of other IO or IP devices.

The enabling technologies for IP mainly depend on silica-on-silicon (SOS), in which the structure of waveguide comprises three layers, named as; the buffer, the core and the cladding. The real benefit of SOS is the ability to apply wafer-scale, planar lithography and processing techniques to integrate substantial numbers of functions either as arrays of identical devices or in the form of customized circuit configurations on single or multiple chips. This integration capability offers an efficient platform for the implementation of typical fiber-based functionalities such as optical power splitters or combiners, couplers, wavelength-selective couplers, multiplexers/demultiplexers and optical gain elements. Furthermore, optical switches and controllable attenuators based on the thermo-optic effect can also be fabricated [2].

The Four Wave Mixing (FWM) in optical fibers can be both useful and harmful, this depends on the application. It can be harmful as it is capable of inducing crosstalk in WDM communication systems and limiting the efficiency of such systems. Nevertheless, FWM can be made avoidable by using asymmetrical channel spacing's or using fibers having bulky enough GVD that the phase of FWM process is not matched over long lengths of fiber.

The FWM is the process that becomes fairly efficient if the phase-matching condition is fulfilled in the sense that the efficiency can be made to go beyond one. From practical point of view, more power is appeared at the new wavelength in comparison with the power of the signal happened to fall at the input end. This is un-surprisingly true if we note that the pump beam provides energy to both the idler wave and signal wave at the same time. The usage of FWM for conversion of wavelength engrossed significant consideration during the 1990s because of its potential application in light-wave wavelength division multiplexing (WDM) systems. If a pump beam along with a pulse train of signals is injected together, that contains a sequence of "1" and "0" bits which is pseudo-random inside a parametric amplifier, the wave is generated as a result through FWM only when the pump and signal are presented after one another. This results the idler wave that appears in the sequence form of a pulse train containing "1" and "0" bits as the signal. In consequence, FWM is capable of transferring the signal data to the idler wave at a wavelength that is new with perfect reliability. It can make a signal even more improved in term of quality by reducing the noise intensity.

The all-optical wavelength conversions (AOWC) is significant enough to be an indispensable technology for the switching architectures of future. There are numerous schemes, such as a cross-phase modulation (XPM), a self-phase modulation also known as SPM, a cross polarization modulation (XPolM), a cross-gain modulation (XGM), can be used to realize the all optical wavelength conversion. Nevertheless, the AOWC is based on a four-wave mixing, one of the most promising schemes, because of the reason that it is fully see-through to the signal modulation format and bit rate of signal.

In optical fibers the processes of scattering mostly depends on molecular vibrations or variations in density of silica. In another category of nonlinear phenomena, optical fibers participate by playing a passive role excluding the mediating interaction among a number of optical waves. The processes that are nonlinear are referred to as parametric processes, it is because of the involvement of medium parameters modulation, such as the refractive index, and phase-matching is required before they can build up along with the fiber. Among these, four-wave mixing (FWM) is a technique that plays a vital role. Although FWM can be unfavorable for Wavelength Division Multiplexing systems that should be designed in such a way that the impact is reduced, it is also functional for a range of applications like generation of a spectrally inverted signal through the optical phase conjugation process, designing light wave systems and wavelength conversion. The FWM can also be made applicable for holographic imaging, phase conjugation, and optical image processing.

FWM can be performed in optical amplifiers like semiconductor optical amplifier (SOA) or fiber, in general, the AOWC is based in fiber on the FWM, it can be operated at a higher speed, while the AOWC speed in the semiconductor optical amplifier is inadequate due to the reason that the response time of the carriers is long. The AOWC is basically based on the FWM with a single pump, has been demonstrated in a high-nonlinear dispersion shifted fiber, but the conversion is sensitive in polarizing aspect, and the spectral in converted form is inverted comparative to that of the original signal. Switching network is a potentially promising scheme of switching in the next routing generation and forwarding an ultrahigh bit-rate data in the optical layer without detecting the payload.

There are numerous schemes in order to reduce the sensitivity of polarization of the FWM-based wavelength conversion by using the dual pumps. A polarization scheme with multiplicity architecture requires an additional optical components as well as an accurate polarization control in the loop. In the case of twisted optical fiber, the dual-pump FWM with circular-polarization pumps and signal can also minimize the sensitivity of polarization of the signal in converted form, the setup although, is complex. The converted signal is a circular polarization, and it has limited conversion bandwidth because of the circular polarization-mode dispersion of the twisted fiber. For the fiber which has standard of tens-kilometer, with a birefringence that is low, the birefringence reorientation changes its axes randomly along the fiber, this eliminates the reliance of the signal polarization. Referring the Manakov equation, Inoue and McKinstrie theoretically revealed that the scheme with orthogonal pumps can minimize the sensitivity of polarization.

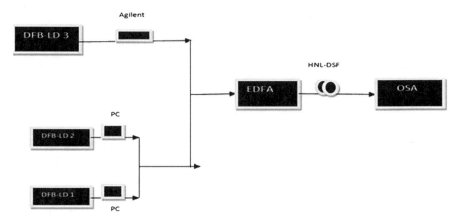

Fig. 1.3 Example of experimental setup of wavelength conversion based on FWM

Refer to Fig. 1.3, a general setup for wavelength conversion is shown where, DFB-LD stands for Distributed Feedback Laser Diode, HNL-DSF referred as High Non-Linear Dispersion Shifted Fiber, EDFA is Erradium Doped Fiber Amplifier, PC is referred as Polarization Controller, and Optical Spectrum Analyzer represented as OSA.

Recently, Silicon photonics is the field that has emerged as a potentially promising technology podium for the low-cost solutions in optical communications and interconnections. Lately, many significant breakthroughs have been reported including amplifiers, silicon modulators, and lasers. All-optical wavelength conversion which is actually based on anti-Stokes Raman scattering that is coherent in nature or four–wave mixing in silicon waveguides is another striking area that has been explored and gaining attention day-by-day. Nevertheless, the efficiency conversion reportedly remains in the range of -50 dB to -35 dB. One of the major restrictions is absorption which is strongly nonlinear occurring at high pump powers. While linear optical absorption in silicon at wavelengths of range lies from 1.3 to 1.7 μm, two photon absorption (TPA) causes increase in optical loss due to induced free carrier absorption (FCA). It has been demonstrated that the TPA induced FCA in silicon can be reduced significantly by introducing a p-i-n diode reverse biased in structure, embedded in a silicon waveguide, reducing the life time of free carriers.

While using such p-i-n waveguides, FWM conversion efficiency achieved is up to -8.5 dB, which is as good as devices based on LiNbO3. High-speed optical data stream at 10 Gb/s on one DWDM channel in the C-band can be made to be converted to some other channel with nearly no distortion in waveform.

The problem faced while light propagating through silicon waveguides having distinguishing lateral dimensions of the order of 1 μm has been widely studied in recent years, both theoretically and experimentally, because of its enormous practical applications. Silicon is considered a promising significant photonics material stems because of its relatively stronger nonlinear interaction with external electromagnetic fields, wavelengths of whose lie in the region of transparent infrared region away from 1.1 μm. Since the telecommunication window near 1.55 μm lies in this region, a multitude of nonlinear optical effects inside silicon waveguides can be used for miscellaneous useful applications. Furthermore, these nonlinear interactions can be improved by employing silicon-on-insulator (SOI) waveguides in which a tight-mode confinement provides large optical intensities even at moderate input power levels. Therefore, it is not surprising that, to-date, almost all physical properties of silicon have found applications in different nonlinear SOI-based photonic devices. Like, stimulated Raman scattering (SRS), which is particularly strong in silicon, is employed to make optical modulators, amplifiers, and Raman lasers.

The Kerr effect is successfully applied for soliton formation, optical phase modulation, and super continuum generation. The phenomenon of four-wave mixing by itself, or in combination with SRS, has been used to make broadband frequency converters. Although two-photon absorption by itself is undesirable, it has been demonstrated that TPA-induced free-carrier generation and thermo optic effects are suitable for all-optical switching, modulation, and pulse compression; they can also be used for autocorrelation measurements. The natural compatibility of SOI technology with the existing silicon manufacturing process opens up wide possibilities for utilizing these and other useful functionalities in fabricating photonic integrated circuits.

To date, nonlinear propagation of optical pulses through silicon waveguides has been studied mostly numerically by using the well-known, split-step Fourier method. It makes use of widely deployed slowly varying envelope approximation to separate a rapidly varying waveform (the carrier) from the signal (the envelope). Another numerical method, which is often used for a direct solution of the Maxwell's equations, is the finite-difference time-domain (FDTD) method. Since it does not make use of the slowly varying envelope approximation, the FDTD scheme is well suited for studying the propagation of pulses as short as a single optical cycle. In principle, these two numerical methods can provide comprehensive information and model all types of nonlinear phenomena inside silicon waveguides.

In spite of this, simple analytical solutions and semi-analytical tools are of considerable value in practice because they offer a clearer view of nonlinear processes in silicon waveguides and may open up nontrivial paths for device optimization. Over the past few years, a number of such methods have been proposed in literature on nonlinear silicon photonics.

For an intuitive understanding of nonlinear optical phenomena in silicon waveguides, numerical simulations should be supported by simple analytical solutions capable of providing a reasonable estimate of the physical quantity involved. Such solutions provide not only a rapid way of checking simulation results but also considerable physical insight that is often lost in voluminous data generated by numerical simulations.

While the impact of communication needs of modern multinational corporations, governments and institutions on rapid development of infrastructure is undisputed, we are at a point where the individuals and their needs is also an important factor in driving the modern communication applications, such as Internet related (e-mail, web), wireless (cell phones) and television etc. All these applications have become a way of life in modern world.

Modern era is all about communication and the communications rely heavily on the ability to transfer the ever increasing amount of information across the globe in the most efficient manner. Networks may be classified according to a wide variety of characteristics such as medium used to transport the data, communications protocol used, scale, topology, organizational scope, etc. Medium is thus of utmost importance, it should be capable of fulfilling the requirements with least complexity and easy troubleshooting, compact in size, flexible and extremely rugged.

Presently, we are living in the information age that has evolved by the exponentially developed and technically enhanced internet technology. The evolution of the internet and pervasive computing creates massive bandwidth demand for data communication. As industries move high to higher bandwidths, electronic communication links are approaching the fundamental distance × bandwidth limitation governed by the physical loss mechanisms (a function of distance and bandwidth) and noise levels. For example, the bandwidth for today's high-speed copper interconnects in computing systems is roughly 3 GHz within a 1 m distance. Most experts feel that copper interconnect limits will be reached in the 10–40 GHz range for PC board materials within 1 m distances meaning that these expected limits will be reached in only a few generations. Moving close to a fundamental limit usually involves ever-increasing costs. To ease cost pressure, many industries have chosen to find alternative technology platforms that do not suffer from the same physical limitations.

For many industries with long communication distances such as the long-haul industry and the storage area network industry; this choice has been to switch to optical fibers [1]. As the communication bandwidth demand gets higher and higher, applications with shorter and shorter communication distances are expected to migrate to photonics. Moreover, the information carrying capacity of optical fiber cable has been proved much greater than any other transmission medium. The utilization of bandwidth can be increased by several times if multiple optical signals are transmitted at the same time.

The graduate level research work that is presented in this book is mainly divided into two main sections. The first section deals with amplification and wavelength conversion whereas the second one describes optical signal processing using nonlinear photonics in optical communication systems.

The further distribution of the work presented in this book, is spelled out as under;

Chapter 2 presents the modern trends and the work carried out in the area of optical signal processing using nonlinear photonics in telecommunication systems. The work spreads over the overview from different publications on optical network management, optical signal processing using nonlinear photonics and major schemes and methodologies exploited in this area over the decades.

Chapter 3 highlights the standard of optical communication and network theory. It also examines the characteristics of optical communication systems by discussing different communication media. The chapter also points out the devices being used and considered for future consideration too.

Chapter 4 focuses on the gradual and time-needed advancements made in the field of photonics. It also highlights the advancements made in the devices which are being used and are being prepared for future trends as well. The traditional and nontraditional motivations regarding selection of silicon as the enabling material for the usefulness of photonics and the types of SOI waveguides are also presented with great emphasis and depth.

Chapter 5 presents the theory and subsequent implementation of high-nonlinearity in glass fibers and cross phase modulation (XPM) whereas Chap. 6 gives detail of the super-continuum generation by nonlinear optics.

Chapter 7 provides a theoretical model for pulse propagation inside an SOI waveguide. The FWM process in SOI waveguides is discussed with an emphasis on the effects of two-photon absorption and the consequent free-carrier effects. All optical wavelength conversion is the most repeatedly requisite phenomenon in present and future wavelength division multiplexing transmission systems. The utilization of multiplexing techniques such as dense wavelength division multiplexing (DWDM) and optical time division multiplexing (OTDM) has shown an enormous increase in the optical transmission capacity.

Chapter 8 describes the FWM and its types in detail. It also depicts the mathematical equations which provide basis for mathematical modeling and subsequent realization on various platforms. For achieving optical frequency shifting by FWM in SOI waveguide, a nonlinear phenomenon has been presented. The FWM process in SOI waveguides is also discussed with an emphasis on the effects of two-photon absorption and the consequent free-carrier effects.

Chapter 9 elucidates the research work carried out on Matlab simulink for optical wavelength conversion by using FWM process in SOI waveguides. It also extracts conclusion of the research and offers suggestions for future attempts in the subject area. The last chapter also deals with the losses due to two photon absorption and free carrier absorption, and how to decrease these losses with SOI waveguides. The summery of this work, future recommendations and the research directions of the authors for near future is mentioned.

References

1. Proakis JG, Salehi M (2002) Communication systems engineering. Prentice Hall, Englewood Cliffs
2. Senior JM (2009) Optical fiber communications principles and practice. Prentice Hall, Englewood Cliffs

Chapter 2
Contemporary and Future Trends in Optics

Abstract In this chapter research based work encompassing present and future trends of optics has been presented. Various research aspects in the area of silica-glass fiber have substantially been explored and pointed out for future trends. An effort has been made to incorporate the latest and updated research so as to provide the premises for the work that has been done in the subject area.

Keywords Continuous wave laser • Optoelectronic integration • Photonic integrated circuits • Nonlinear interaction • Signal to noise ratio

Abbreviations

LED	Light-emitting diodes
PICs	Photonic integrated circuits
CW	Continuous wave
SRS	Stimulated Raman scattering
TPA	Two photon absorption
CPM	Cross phase modulation
FCA	Free carrier absorption
GVD	Group velocity dispersion
NF	Noise figure
OSNR	Optical signal to noise ratio

The invention of the silica-glass optical fiber having loss range acceptable for information transportation using chemical vapor deposition, facilitated new possibility in communication technology, and led to possibility of ultra-high capacity information flow. Optical age begins with two inventions, almost concurrently in 1970s, low loss silica-glass fiber and continuous wave (CW) semiconductor laser.

J. Ahmed et al., *Optical Signal Processing by Silicon Photonics*, SpringerBriefs in Materials, 11
DOI: 10.1007/978-981-4560-11-5_2, © The Author(s) 2013

Both of these devices were able to operate at room temperature, although the life of laser was short enough and was addressed immediately.

The high cost of initial fiber was soon declined by the promising ability of excellent transmission properties and high bandwidth. Audio and video transmission applications for broadband systems forced optical manufacturers to develop high volume mass production facilities of fibers which consequently lower the cost of the fiber. Dr. Charles Kao who presented optical communications concept back in 1970s, let him won the Nobel prize in 2009. In the start of this chapter the basic principle of optical communication is presented followed by the recent advances with applications in the field of silicon photonics.

Soref [1] provides an overview of the past, present, and future of silicon photonics. The focus in the work already carried out is on the main components and integration of optoelectronic components. Identification of the forthcoming challenges and emerging trends has been attempted to be done. The work in this particular area represents kind of revolution in the field of photonics. Covered topics include fast modulators, Raman lasers, light-emitting diodes (LEDs), Quantum-cascade structures, photo detectors, micro resonators, Plasmon optics, photonic-circuit integration, all of these topics are touched briefly, and a few words have been added about the quest for electric field effect modulators, electrically pumped silicon lasers and photonic integrated circuits (PICs) for the long-wave, mid-wave, and far-infrared regions. The text not only points out key references but also provide valuable background to the beginners.

Singh and Singh [2], gives a brief account of their work i.e. to nonlinear effects, for example Self Phase Modulation, Cross Phase Modulation, and Four Wave Mixing have been discussed. These effects cause the degradation of fiber optic system's performance. SPM has tiny impact if power per channel is below 19.6 mW. Four wave mixing is the phenomenon that has rigorous effects in Wavelength Division Multiplex systems, dispersion-shifted fibers are used in it. In case of some dispersion, the effects of four wave mixing are reduced. The non-zero dispersion-shifted fibers are very efficient in WDM systems for this very reason. Besides the degradation effect, they are also useful in various applications including Cross Phase Modulation in optical switching, Self Phase Modulation in solitons, and Four Wave Mixing in wavelength conversion and squeezing (Fig. 2.1, Table 2.1).

Lin and Agrawal [3], gives a brief account of their work that nonlinear optical effects and their several kinds have been under observation recently making use of silicon waveguides, and their applications in different devices are attracting the attention considerably. In this review, they offer an integrated theoretical platform that is capable of understanding the physics underlying and providing the required guidance toward latest and valuable applications. They initiate with the explanation of silicon's third-order nonlinearity with the inclusion of the tonsorial nature of both the Raman and electronic contributions. Free carriers generation through absorption of two-photon and their influence on several nonlinear phenomena is fully included within the presented theory. The derivation of a general propagation equation is done in the frequency domain and how it leads to a generalized

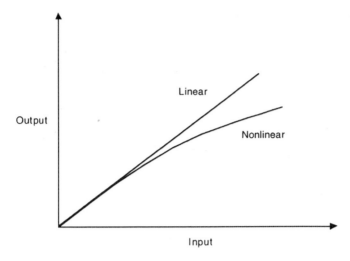

Fig. 2.1 Linear and nonlinear interactions

Table 2.1 Comparison of nonlinear effects

Characteristics	Nonlinear phenomena		
	Self-phase modulation	Cross-phase modulation	Four-wave mixing
1. Bit-Rate	Dependent	Dependent	Dependent
2. Origin	Nonlinear susceptibility	Nonlinear susceptibility	Nonlinear susceptibility
3. Shape of broadening	Symmetrical	May be symmetrical	–
4. Energy transfer between medium and optical pulse	No	No	No
5. Channel Spacing	No effect	Increases on decreasing the spaces	Increases on decreasing the spaces

nonlinear Schrodinger equation is shown when its time domain conversion is done. The four-wave mixing process for both pulsed and continuous-wave pumping is considered and discussed the circumstances under which wavelength conversion and parametric amplification can be realized with possible gain in the telecommunication band.

The nonlinear phenomena of cross phase modulation and SRS were discussed emphasizing on the effect of free carriers on lasing and Raman amplification. The main focus of their work was on the Four Wave-Mixing process and its applications. They also took into account the impact of free carriers first and showed that, besides the index changes induced by them caused a insignificant impact on Four Wave-Mixing, efficiency of FWM is effected by FCA such a way that considering the CW pumping case, a net positive gain cannot be realized in the telecommunication band.

As a solution to this problem pumping is done at wavelengths beyond 2.2 μm due to the reason that the TPA-induced free carriers are absent. Four Wave-Mixing can take place over a wide bandwidth, by making a proper pump wavelength choice, due to the reason that the waveguide lengths are much smaller, they showed. The use of four wave-mixing in silicon waveguides is elaborated sufficiently, for generating correlated photon pairs that are useful for quantum applications.

Foster et al. [4] used phase-matched FWM in properly designed silicon waveguides in order to describe 29 nm range amplification and conversion of wavelength in efficient manner, ranging between 1,511 and 1,591 nm. The crucial characteristic of their work is the design of the waveguides suitably in order to fabricate irregular group-velocity dispersion in this management of wavelength. Unlike the Raman Effect in silicon, the supple FWM's pump-signal detuning, allows both the signal and pump to exist in the communications band. Through parametric wavelength conversion and optical parametric oscillation, this advancement makes possible the implementation of from a single pump laser in all-silicon photonic integrated circuit. Moreover, all-optical switches, all optical delays, optical sources and optical signal regenerators for quantum information technology, which have been realized using four wave-mixing in silica fibers, can be ported to the SOI platform. FWM amplification depends critically on the phase mismatch between the pump, signal and idler waves.

Liang et al. [5], discussed in their research the conventional silicon-based optical switching device's switching speed, based on plasma dispersion effect is restricted by the lifetime of free carriers which introduce either absorption or phase changes. They neither state all-optical NOR gate logic which is independent of free carriers but rely on two-photon absorption. Pump induced non-degenerate two-photon absorption causes high speed operation, inside the silicon wire waveguides of submicron size. For logic gate operation, the device needed low pulse energy (few pJ).

Lee et al. [6], carried out work and showed ultra-broadband wavelength conversion in silicon photonic waveguides at 40 Gb/s data rate. The device which is built in such a way that they avoid dispersion, elaborates a bandwidth conversion spanning the entire L-, C-, and S-bands of the ITU grid. For using a continuous-wave -band pump, the 1,513.7 nm wavelength input signal is up-converted across nearly 50 nm with 40 Gb/s data rate, and bit-error rate measurements are performed on the converted signal.

They selected a Dense Wavelength-Division Multiplexing operation on the ITU—band's channel C50 to couple the probe beams and pump. The wavelengths of probe and pump were chosen for providing conversion bandwidth which is the largest bandwidth that can be allowed by the tunable laser and tunable filters. The trace of the sample OSA shows the 47.7 nm conversion from 1,513.7 nm wavelength input probe to a converted 1,561.4 nm wavelength of with conversion efficiency near 18 dB having 40 Gb/s data rate. A visible degradation in optical signal-to-noise ratio of converted signal can be observed due to the conversion losses. For this reason, the efficiency of the wavelength converter is one of its most crucial parameters for systems-level integration.

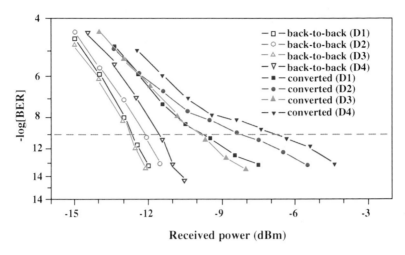

Fig. 2.2   The BER curve of 40 Gb/s shown on each output of the demultiplexer (D1–D4) [7]

Figure 2.2 shows BER curves of the converted signal at the rate of 40 Gb/s for each output of the four de-multiplexer (D1–D4). After the modulation the back-to-back curve is taken directly (no amplifiers, chips or filters in the optical pathway) of a probe signal which is set to the same wavelength as that of the signal which is converted. They described the high-speed longest-spanning wavelength conversion reportedly lie in a dispersion engineered silicon photonic waveguide, and have characterized the BER degradation quantitatively produced as a result from the conversion.

The degradation cause is determined primarily and it is found to be due to the conversion efficiencies which are limited, but in near future some improvements are really expected in active devices. Thus, the results provide advancement significantly toward the ultrafast all-optical parametric processing devices integration within large-scale optical networking systems.

Turner-Foster et al. [7] demonstrated ultra-broadband low-peak-power frequency conversion of CW light in a silicon photonic structure via four-wave mixing. The process produces continuous conversion over two-thirds of an octave from 1,241- to 2,078-nm wavelength light. Also investigate FWM in silicon waveguides with the zero GVD (group-velocity dispersion), point near the center of the C-band where high power pump lasers are readily available. This design is critical to enable efficient broadband parametric conversion. Pumping near this zero GVD point and observed extremely broad conversion bandwidth with over 830 nm allowing for conversion from 1,241 to 2,078 nm continuously.

The signal and pump are launched into the waveguide's TE-like mode all the way through the inverse nano-taper. Signal is then scanned from 1,241 nm to the C band's pump wavelength. The average power of the signal and pump are 110 and 1.1 mW respectively, in the waveguide. The bandwidth of conversion very much rely on the pump wavelength location with respect to the waveguide's zero GVD

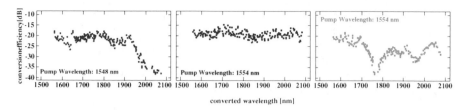

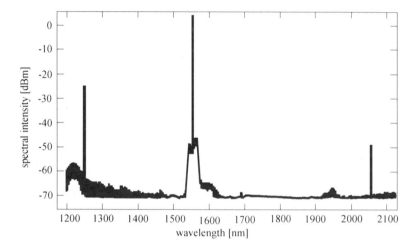

Fig. 2.3 Experimentally measured conversion efficiency as converted wavelength's function of 1,548, 1,554, and 1,560 nm (from *left* to *right*) [7]

Fig. 2.4 Sample showing four wave-mixing plot of converted signal light from 1,250 to 2,056 nm wavelength where 1,554.7 nm is the pump's wavelength with −24 dB of on–off conversion efficiency [7]

point and careful tuning of the pump wavelength allows for optimization of the conversion bandwidth. The zero GVD point for this waveguide tracked at a wavelength of 1,554 nm, and the dispersion slope is 1.9 ps/nm^2 km. The conversion bandwidth is extended to 1,930 nm experimentally and then there is a considerable decrease for a short pump wavelength as shown in Fig. 2.3 left plot. In case of a larger pump wavelength of 1,560 nm, the initial bandwidth converted experimentally increased up to 1,760 nm only, the oscillatory features are really sharp as demonstrated in the rightmost trace of Fig. 2.3.

However, placing the pump at the zero GVD wavelength of 1,554 nm, the conversion efficiency is maintained until 2,078 nm where we reach the tuning limit of our signal laser (1,241 nm). The conversion efficiency over this bandwidth is around −18 dB and can be seen in the middle trace of Fig. 2.3.

Figure 2.4 shows the wavelength conversion from 1,250 to 2,056 nm when the pump is centered at 1,554.7 nm, the figure further demonstrates about −24 dB on–off conversion efficiency and over 830 nm of shift. The ability to generate and convert wavelengths over such a large range of tuning ability increases silicon photonics' functionality by adding new functions including on-chip spectroscopy, high data rate on-chip signal processing, on-chip ultrafast pulse measurements, and high speed telecommunications.

References

1. Soref R (2006) The past, present, and future of silicon photonics. IEEE J Sel Top Quantum Electron 12(6):1678–1687
2. Singh SP, Singh N (2007) Nonlinear effects in optical fibers: origin, management and applications. Prog Electromagnetics Res PIER 73:249–275
3. Lin Q, Painter OJ, Agrawal GP (2007) Nonlinear optical phenomena in silicon waveguides: modeling and applications. Opt Exp 15(25):16604–16644
4. Foster MA, Sharping JE, Gaeta AL (2006) Broad-bandwidth optical gain and efficient wavelength conversion in silicon waveguide. 1-55752-8132006 IEEE
5. Liang TK, Nunes LR, Tsuchiya M, Abedin KS, Miyazaki T, Van Thourhout D, Bogaerts W, Dumon P, Baets R, Tsang HK (2006) High speed logic gate using two-photon absorption in silicon waveguides. Optics Commun 265:171–174
6. Lee BG, Biberman A, Turner-Foster AC, Foster MA, Lipson M, Gaeta AL, Bergman K (2009) Demonstration of broadband wavelength conversion At 40 Gb/S in silicon waveguides. IEEE Photonics Technol Lett 21(3):182–184
7. Turner-Foster AC, Foster MA, Salem R, Gaeta AL, Lipson M (2010) Frequency conversion over two-thirds of an octave in silicon nanowaveguides. Optics Express 18(3):1904–1908

Chapter 3
Modern Optical Networks

Abstract In this chapter, the standards of optical communication and network theory are highlighted with their applications and significance in the respective areas. It also examines the characteristics of optical communication systems by discussing different communication media. The devices being used and considered for future consideration are also discussed and summarized for optical networks.

Keywords Optical networks • All-optical networks • Regeneration of optical signals

Abbreviations

LAN   Local area network
WAN   Wide area network
MAN   Metropolitan area network
WDM   Wavelength division multiplexing
HNL   Highly nonlinear
SLD   Semiconductor laser diode
SHG   Second harmonic generation
OEO   Optical signal is converted to electrical signal
MMF   Multimode fibers

The first generation optical networks were essentially point-to-point links where the fiber-optic link was used only as a replacement for the copper cables [1]. The optical signal propagating through the optical fiber was detected electronically, regenerated using a semiconductor transponder, and transmitted optically over fiber to the next opto-electronic repeater. The routing and switching functions were done electronically at the network nodes. In addition, each node had to handle not only the traffic destined for it but also all other traffic that was being routed through that node.

J. Ahmed et al., *Optical Signal Processing by Silicon Photonics*, SpringerBriefs in Materials,
DOI: 10.1007/978-981-4560-11-5_3, © The Author(s) 2013

One of the first commercial steps toward an all-optical network was taken in 1988 when the first optical fiber was installed between Europe and America. Since then the telecommunication industry has been through a major economic boom and a crash and is now on the road to recovery. Since the first installation of optical fibers between the continents several companies and network providers have installed more fibers leading to a massive increase in capacity. With the increased internet traffic through the 1990s the demand for more capacity kept rising and network providers kept delivering. The bandwidth was supplied with the aid of the newest advances in research such as wavelength division multiplexing (WDM).

With the increase of transmission rates to 10 Gbps and higher, the fiber medium impairments like dispersion and nonlinearities [2] start to emerge. On the other hand, the electronic processing of every packet at a node becomes more challenging in terms of requirements for electronics, buffering and power consumption, and processing power. Therefore, the second generation optical networks have been created to incorporate some of the switching and routing functions in the optical layer of the network [3].

The main approach in realizing the second generation optical networks is frequency division multiplexing in the optical domain. This concept is utilized in electrical communications to increase the communication channel capacity. Optical frequency division multiplexing involves combining data transmitted on different optical wavelengths within the same fiber, thus it is most commonly referred to as WDM. It offers a potential for very effective use of fiber bandwidth directly in the time domain [4] while allowing for using the wavelength to perform functions like routing and switching. WDM networks have been deployed in local metropolitan area networks (MAN) and local area networks (LAN) over the past few years. Currently, WDM networks are using hundreds of wavelengths operating at a bit rate of 10 Gbps each. All the routing and switching is still done electronically and the cost of converting signals from the optical to the electrical domain and back again is expected to increase as the bit rate per channel increases. Physical size increase, thermal dissipation and power consumption increase with increasing bandwidth are other major limiting factors of electronic switching and routing [5]. In addition, it is understood that as transmission rates continue to grow, the speed of the electronic circuitry will not be able to compete with the speed of the optical transmission.

In the next generation optical networks [6], it is expected that all-optical switches will be used to perform switching and routing functions in the optical domain. The major advantage to this approach is the tremendous bandwidth potential. The all-optical approach may lead to potential lowering of cost per bandwidth, but it will most certainly help with reduction of the footprint of the gear and its power consumption and dissipation issues. This is contingent on development of photonic integrated circuits, which are small in size, consume low energy and capable of switching large bandwidths of data very fast.

3.1 Evolution of Optical Fiber Communication Systems

The principles of fiber-optical communication systems are covered in many textbooks, for example, [7, 8]. This section describes the basic principle of optical network. Optical communication link is shown in Fig. 3.1. Four key elements make up nearly every optical communication link. The transmitter at the input of link consists of a light source, typically a semiconductor laser diode (SLD) since the SLD gives longer transmission distances than other light sources. The light is modulated either directly by varying the laser driving current, or by an external modulator. At the output of the link a photodetector converts the optical data signal into an electronic signal [9]. The transmission medium for the optical data signal is a fiber-optic cable.

The fourth element of link is the regenerator. For a long transmission distances, regenerators can be used between the transmitter and the receiver to remove the noise and recover the signal distorted along the fiber. The regenerators used for single channel system are generally OEO regenerator, in which the optical signal is received (optical signal is converted to electrical signal), regenerated, and re-transmitted by a transmitter (electrical signal is converted back to optical signal). OEO converters are also format-dependent, and their speed is limited by available electronics. Nevertheless, OEO converters are so far one of the most mature technologies, due to its straight forward approach.

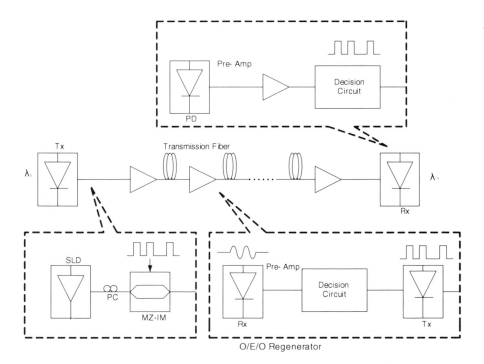

Fig. 3.1 A typical optical communication link [8]

In the search of achieving all-optical components such as wavelength conversion and regenerators various methods have been used. One of the ways of obtaining all-optical signal processing has been by using nonlinear effects.

3.2 Regeneration of High Speed Optical Signals

Regeneration is an integral part of a data transmission system and therefore a lot of effort has gone into research on regeneration, and in this context especially all-optical regeneration has been a research focus in recent years. Regeneration consists of three parts: Re-amplification, reshaping and re-timing of the data signal, where the first two ways combine to 2R-regeneration and including re-timing 3R-regeneration is achieved.

Regeneration of an optical signal can be performed either all-optically or optoelectronically. Opto-electronic regeneration consists of three parts: detecting the optical signal, regenerating it electronically, and using this electronic signal to modulate a new optical carrier. The latest advances in the research on opto-electronic regeneration have increased the speed of opto-electronic regenerators to bit rates up to 40 Gbit/s [10].

Opto-electronic 3R-regeneration is a simplification over all-optical regeneration in terms of parameters that needs to be controlled. The implementation of opto-electronic 3R-regeneration in systems today is generally preferred by system providers over all-optical 3R-regeneration [11].

The choice of one technology over the other is not a simple one. Since regenerators will be implemented in future WDM systems, the number of regenerators increases with the number of channels in a WDM system. That means that besides the number of controllable parameters, also cost, number of components, and power budget of the regenerator need to be considered. This is a complex task that can only be decided on a case by case basis.

For signals with higher bit rates of 80 or 160 Gbit/s all-optical regeneration is the only option. Therefore all-optical regeneration remains an important research tool and much investigation is put into that area of research. Some of the latest results using SOAs [12, 13] or fiber in a 3R-regenerator shows a promising future for all-optical 3R-regeneration [14]. An all-optical regenerator is a device that receives a noisy and distorted signal and transmits an undistorted and noise free signal.

The loss in a fiber is also wavelength-dependent. Different frequency components of the propagating light are attenuated with different magnitudes. The factors which contribute to the loss spectrum are the Rayleigh scattering, water (OH^-) absorption and metal-oxide absorption peaks. Silica glass, for example, has electronic resonances in the UV region.

Therefore, this type of glasses can transmit light in the 0.5–2.2 μm region. Rayleigh scattering is a fundamental loss mechanism arising from the density fluctuations frozen into the fused silica during manufacturing. Resulting local

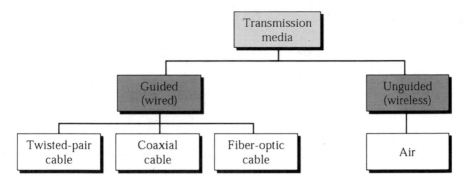

Fig. 3.2   Transmission media classification

fluctuations in the refractive index scatter light in all directions. Another loss factor is the bending loss which may scatter light at the core-cladding interface. In communication systems the splicing loss and connector losses may also contribute to the attenuation of the transmitted light (Fig. 3.2).

3.3 Transmission Media

- Guided media: that provides a medium from one device to another. For example twisted-pair, coaxial cable, and optical fiber.
- Unguided media/wireless communication: that transfers electromagnetic waves without using a physical conductor. As a substitute, signals are broadcast through air (or, in a few cases, water), and thus are offered to anyone who has a device capable of receiving them.

3.4 Communication Media

The communication media is briefly divided as under on the basis of the physical structure required for it.

3.4.1 Wired Technologies

Twisted-pair is generally the most common medium used for telecommunication. In twisted-pair cabling, the copper wires are twisted to form pairs, the typical telephone wires consist of two insulated copper wires twisted into pairs. In computer

Fig. 3.3 A cross-sectional
view of optical fiber [16]

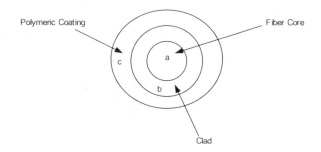

networking, cabling can be utilized equally for voice and data transmission, for
this it consists of 4 pairs of copper cabling. The two wires, twisted together can
reduce electromagnetic induction and crosstalk effectively. The speed of transmission range starts from 2 million bits per second to 10 billion bits per second.

Coaxial cable is generally used in cable television systems, in office build-
ings, and work-places for LANs. The cables are formed using aluminum or copper wire enfolded in insulating layer typically of a material with a high dielectric
constant and flexible in nature, finally all of it is enclosed in a conductive layer.
The layers of insulation help minimize interference and distortion to a great deal.
Transmission speed ranges between 200 million to more than 500 million bits per
second.

Optical Fiber Cable one or more glass fiber filaments are wrapped in protective layers. It also transfers data by means of light pulses. It carries light which
can travel over the extensive distances. Fiber-optic cables are unaffected by the
electromagnetic radiation. The transmission speed might cross trillions of bits per
second.

The cross-sectional view of an optical fiber is shown in Fig. 3.3. Optical fiber
consists of core, clad, and protective polymer coating. Fiber core and clad are
made up of glass and the optical fibers with core. The diameters of fiber is greater
than 10 μ and known as multimode fibers, moreover the fibers with core diameters
less than 10 μ microns are known as single mode (Fig. 3.4).

The transmission speed of fiber optics is hundreds of times quicker than that
for coaxial cables and in comparison to twisted-pair, the transmission speed of
fiber optics is thousands of times quicker than a twisted-pair wire. So the usage
of colored light might increase this capacity further, i.e., light of multiple wavelengths. As a substitute of transportation of one message in a stream of monochro-
matic light impulses, this technology is capable of carrying multiple signals in a
single fiber. The most significant advantages presented by fiber-optic cable over
twisted-pair and coaxial cable are noise resistance, less signal attenuation, and
higher bandwidth.

The simplest form of an optical fiber consists of a core of silica glass cylindri-
cal in shape bordered by a cladding having refractive index lower than that of core.
Some other types of fibers are step index fiber and graded-index fiber.

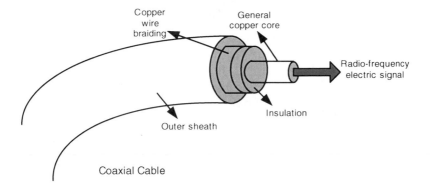

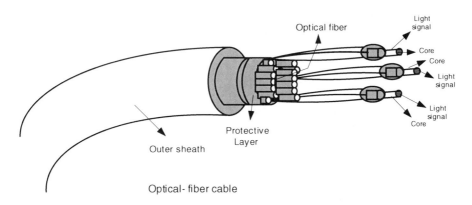

Fig. 3.4   Optical fiber cable

The function of the optical cable may be summarized into four main areas;

• Fiber protection against fiber damage and breakage during installation throughout the life of the fiber.
• Cable strength that may be enhanced by incorporating an appropriate strength member and by providing the cable a thoroughly designed thick outer sheath.
• Stability in the context of the fiber transmission characteristics.
• Identification and joining of the fibers at joints, within the cable. That is very significant for cables having large number of optical fibers.

Standard fibers for communication purposes are step-index or graded-index fibers manufactured nowadays. Graded-index fibers have a core with decreasing refractive index from the center to the core boundary (Fig. 3.5).

These fibers are usually referred to transmission fibers. To increase the transmission capacity and the wavelength range where these fibers are able to be used, dispersion compensating and active fibers were introduced. Considering the

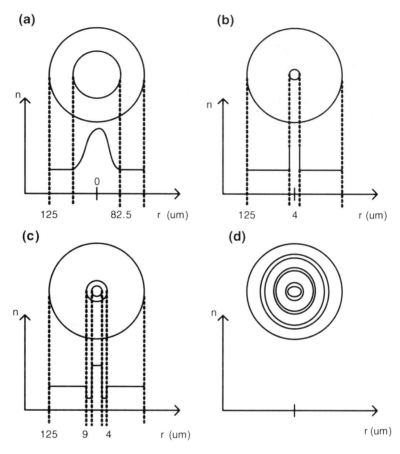

Fig. 3.5 Some typical index profile for optical fibers **a** graded index, **b** step index, and **c** dispersion compensating fiber [8]

geometrical properties and the number of guided modes the fibers can be categorized further as multimode and single mode fibers.

Multimode fibers (MMF) are inexpensive solutions for building LAN in an office, for example. The name MMF comes from the fact that the light travels down the fiber in multiple paths. Graded-index fibers are usually used for LAN/WAN equipments because the light path in this case is sinuous and regular while in the case of step-index MMF the light path is highly angular and irregular. From the end of 1990s, highly nonlinear (HNL) fibers were developed for super continuum generation.

Ethernet: is defined by IEEE 802 and consists of a range of standards and media that aid the communication between devices. Wireless LAN technology is designed to connect devices exclusive of wiring. These devices make use of radio waves or infrared signals as a transmission medium [15] (Fig. 3.6).

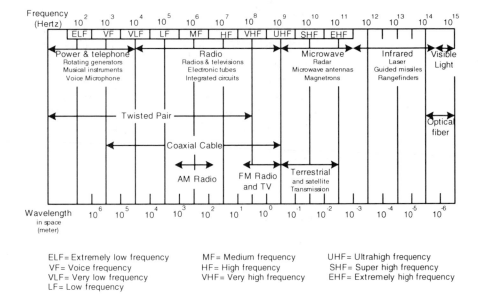

ELF= Extremely low frequency   MF= Medium frequency   UHF= Ultrahigh frequency
VF= Voice frequency   HF= High frequency   SHF= Super high frequency
VLF= Very low frequency   VHF= Very high frequency   EHF= Extremely high frequency
LF= Low frequency

Fig. 3.6   Complete spectrum of electromagnetic radiation

3.5   Dispersion in Optics

In a prism, different colors are refracted at different angles as a result of material dispersion (a wavelength-dependent refractive index) actually it splits the white light forming a rainbow. Dispersion is defined in optics as the phenomenon in which a wave's phase velocity depends on its frequency. The media having such a property are express as dispersive media.

A rainbow is the most renowned example of dispersion in which spatial separation of white light is caused by dispersion, splitting the white light into components of varying wavelengths (different colors). However, dispersion also has an effect in many other conditions: for example, it degrades signals as the distance increases, causing pulses to spread in optical fibers also, a termination between dispersion and nonlinear effects cause formation of soliton waves.

Dispersion is mostly taken into consideration for light waves, but it may occur for any kind of wave that can pass through geometry, inhomogeneous in nature such as sound waves, or any other wave interacting with a medium. Dispersion is at times said to be the chromatic dispersion because it highlights its wavelength-dependent nature.

There are generally two types of dispersion:

1. Material dispersion.
2. Waveguide dispersion.

Material dispersion is caused by the frequency-dependent response of a material to waves. For example, material dispersion that leads to chromatic aberration, which is undesirable in a lens or the separation of colors in a prism.

Waveguide dispersion is that type of dispersion which results when the speed of a wave in a waveguide (such as an optical fiber) is totally dependent on its frequency for geometric reasons, it however does not rely on any frequency dependence of the materials from which it is build. More generally, "waveguide" dispersion can arise for waves that are propagating through any structure, inhomogeneous in nature (e.g., a photonic crystal), whether or not the waves are limited to some section. Generally, both types of dispersion may be present, although they are not firmly additive. Their combination leads to degradation of the signal in optical fibers in telecommunications, because the unstable delay in the time of arrival between different components of a signal "smears out" the in time signal.

3.6 Material Dispersion in Optics

The material dispersion can be a striking or undesirable effect in optical applications. The dispersion of light by glass prisms is utilized in the construction of spectrometers and spectro-radiometers.

3.7 Group and Phase Velocity

Group and phase velocity is the result of dispersion marking itself as a temporal effect. The formula given, $v = c/n$ calculates the phase velocity of a wave; this is the velocity at which the phase of any one frequency component of the wave will propagate. This is not similar to the group velocity of the wave, which is the rate at which the changes in amplitude (known as the envelope of the wave) will propagate. The group velocity 'vg' is associated to phase velocity for a homogeneous medium as:

$$v_g = c\,(n - \lambda \mathrm{d}n/\mathrm{d}y)^{-1}$$

The group velocity v_g is often taken into account as the velocity at which energy or information is transmitted along the wave. In the majority cases this is based on fact, and the group velocity can be considered as the signal velocity of the waveform.

3.8 Optical Frequency Multiplier

An optical frequency multiplier is generally a nonlinear optical device, in this photons are in interaction with a nonlinear material and are effectively "combined" forming new photons which are greater in energy, and thus have higher frequency.

There are two types of devices that are currently more frequent; frequency doublers are generally made up of lithium niobate (LN), potassium titanyl phosphate (KTP), lithium tantalate (LT), or lithium triborate (LBO), and frequency triplers usually are made of potassium dihydrogen phosphate (PTP) or KTP. The two of them are extensively usable in optical experiments that include lasers as a light source.

Two processes are frequently used to attain the conversion, second harmonic generation (SHG) sometimes also known as frequency doubling, or summing of frequency generation which sums two different frequencies. Direct third harmonic generation (THG), also called frequency tripling exists, but famous examples of these devices are less efficient and are generally not used in practice. "Real world" triplers use a two-stage process that combines two photons using SHG, and then addition of a third using summing.

Optical frequency multipliers are more frequent in high-power lasers, markedly those that are used for inertial confinement fusion (ICF) experiments.

3.9 Optical Network

A network can be defined as a collection of transmission links and other utensils which provides a means of information interchange within a group of end users.

The purpose of the network is to provide a means of information interchange between end users. The network usually contains some shared resources. That is, links and switching nodes are shared between many users. The term "network" also usually implies a geographic separation between end users. This is not always true in the sense that communicating end users may be across the room or across the world.

Networks may be characterized by their geographic extent such as:

- Local Area Network (LAN).
- Metropolitan Area Network (MAN).
- Wide Area Network (WAN).

Figure 3.7 illustrates the main concept of a network that is subdivided into three parts;

- Access/edge-network (used to provide the connectivity to user to the nearest node, 1–10 km distances)
- Regional/metro optical network (provide interconnectivity between regional and metropolitan domains)
- Long-haul optical networks (ranges about 10–500 km)

Optical communication networks can support high bandwidth as the WDM techniques are mature enough. The optical industry must observe the different qualities and types of service enabled by different infrastructure configurations and then decide which services to offer over those infrastructures. The optical network can be considered from various geographical domains, multiplexing technologies and transport capacity [16] (Fig. 3.8).

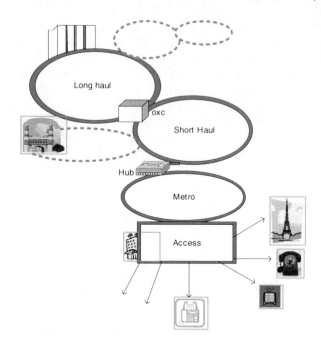

Fig. 3.7 A general view of optical network (distribution, access, and long-haul subnet) [16]

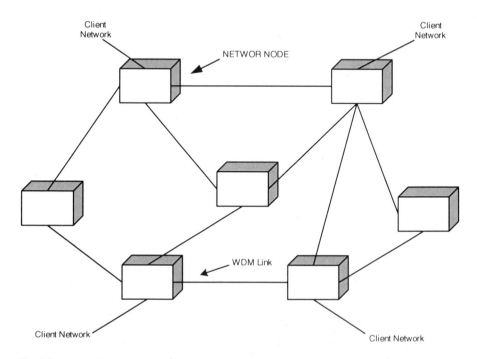

Fig. 3.8 A wavelength routed WDM network [16]

The WDM network presented the routing nodes interconnected by point-to-point WDM fiber link in a meshed topology. In this network, an access device is used for an electronic interface between the optical networks and is client networks can be connected to each node.

Over time we can expect new types of networks to evolve which may have some similarities with electronic networks but which will make use of the properties and characteristics of unique optical components [17].

References

1. Ramaswami R, Sivarajan KN (1998) Optical networks: a practical perspective. Morgan Kaufmann, California
2. Agrawal GP (1997) Fiber-optical communication systems. Wiley-Interscience, New York
3. Ramaswami R, Sivarajan KN, Sasaki GH Optical networks: a practical perspective, 3rd edn
4. Brackett CA (1990) Dense wavelength division multiplexing networks: principles and applications. IEEE J Sel Areas Commun 8:948–964
5. Potasek MJ (2002) All-optical switching for high bandwidth optical networks. Opt Netw Mag 3:30–43
6. Black U (2002) Optical networks-third generation transport systems. Prentice Hall, New York
7. Kazovsky LG, Benedetto S, Willner A (1996) Optical fiber communication systems. Artech house,/norwood, MA
8. Agrawal GP (1997) Fiber-optic communication systems, 2nd edn. Wiley, New York
9. Ramaswami R, Sivarajan KN (2002) Optical networks: a practical perspective, 2nd edn. Morgan Kaufmann, San Franciso
10. Kuebart W, Lavigne B, Witte M, Veith G, Leclerc O (2003) 40 Gb/s transmission over 80,000 km dispersion shifted fibre using compact opto-electronic-3R regeneration. In: Proceedings European conference on optical communication, ECOC'03, Rimini, Italy, p Mo. 4.3.1
11. Nielsen ML (2004) Experimental and theoretical investigation of semiconductor optical amplifier (SOA) based all-optical switches. Ph.D. dissertation, Research Center COM, Technical University of Denmark
12. Lavigne B, Balmefrezol E, Brindel P, Dagens B, Brenot R, Pierre L, Moncelet JL, Grandi`ere DDL, Remy J-C, Bouley J-C, Thedrez B, Leclerc O Low input power all-optical 3R regenerator based on SOA devices for 42.66 Ggit/s ULH WDM
13. RZ transmission with 23 dB span loss and all-EDFA amplification. In: technical digest optical fiber communication conference, OFC'03, Atlanta, Georgia, U.S.A., pp PD 15–1
14. Lavigne B, Balmefrezol E, Brindel P, Pierre L, Dagens B, Brenot R, Thedrez B, Renaud M, Leclerc O (2003) Operation margins of a SOA-based 3R regenerator for 42.66 Gbit/s ULH transmission systems. In: Proceedings European conference on optical communication, ECOC'03, Rimini, Italy, p Mo. 4.3.4
15. Schubert C, Ludwig R, Watanabe S, Futami F, Schmidt C, Berger J, Boerner C, Ferber S, Weber H (2002) 160 gbit/s wavelength converter with 3R-regenerating capability. Electron Lett 38(16):903–904
16. Eelectronic communication system, 5th edn. Wayne Tomasi
17. (2003) The handbook of optical communication networks. Mohammad Ilyas, Hussein T. Mouftah
18. Telecommunications essentials, 2nd edn. Lillian goleniewski
19. Optical fiber communications principles and practice, 3rd edn. John M. Senior
20. (1998) Understanding optical communications. Harry J. R. Dutton

Chapter 4
Advancements in Silicon Photonics

Abstract This chapter focuses on the gradual and time-needed advancements made in the field of photonics. It also highlights the evolution taken place in the devices which are being used and prepared for future trends as well. The traditional and nontraditional motivations regarding selection of silicon as the enabling material for the usefulness of photonics and SOI waveguides and their types are also presented with great emphasis and depth.

Keywords Silicon photonics • Silicon-on-insulator waveguides • Waveguides for silicon photonics

Abbreviations

SOI	Silicon on insulator
CMOS	Complementary metal–oxide–semiconductor
PhCs	Photonic crystal
PIC	Photonic integrated circuit
NLO	Nonlinear optics
ULSI	Ultra-large-scale integrated
IC	Integrated circuit
MEMS	Micro-electro-mechanical structures
MOEMS	Micro-opto-electro-mechanical structures

Reviewing the silicon photonics technology, it is evident that this field has grown rapidly in the recent years. The revolutionized advancement in information technology has spread the interest of nonlinear photonics to every corner of the world. Presently installed power-hungry infrastructure of telecom and information industry has to be improved or changed with silicon-based energy efficient green photonics. The data-hungry new telecom applications, super computer interconnection structures, sensors, and medicine are a few examples of ever increasing

J. Ahmed et al., *Optical Signal Processing by Silicon Photonics*, SpringerBriefs in Materials, 33
DOI: 10.1007/978-981-4560-11-5_4, © The Author(s) 2013

application industries of silicon photonics. The speed of technological improvements is so fast that a new type of Moore's law is needed in silicon photonics industry.

SOI based novel technology enables fabrication of low loss waveguides, high-quality resonators, high speed modulators, efficient couplers, photo-detectors, and optically pumped lasers. Developing efficient detectors and lasers for telecom wavelengths at low-cost are the two main technological hurdles before silicon can become a broad platform for integrated optics. The future of silicon photonics lies mainly in producing very low loss waveguides, integrated designs, spot-sized low loss (less than 0.1 dB) converters, reduced-bend size of resonators, and highly efficient interconnections. The further advancement in photonics industry will be the polarization-independent nature of these devices and overcoming the thermal constraints.

The losses can be reduced by eliminating rough side wall surfaces and careful selection of the waveguide geometries with wavelength selection. Two-photon absorption can potentially be used to avoid the limits and building a better silicon light detection devices on the chip. The fields which are to be concentrated more in future are level of integration, control signaling, and packaging structures. The greater functionality with lower cost can be achieved if the CMOS foundries concentrate more and come up with new design techniques of integration.

4.1 Silicon-on-Insulator Waveguides and Coupling Devices

Silicon on Insulator (SOI) waveguides are attracting considerable attention in recent years, and the devices based on these optical waveguides are important to realize many modern all-optical signal processing systems. The main driving forces behind these devices are the technological improvements in exploiting the nonlinear optical phenomenon and their compatibility with CMOS devices. Although all-optical signal processing devices have been investigated intensively, the chip-scale solution provided by the silicon photonics is the most promising one. Silicon photonic devices are finding their applications not only in telecommunication transmission systems and biomedical instruments, but also in the new generation of high performance computing system.

In recent years, silicon waveguides have shown numerous nonlinear optical applications in various areas particularly in communication. The considerable attention toward the silicon photonics keeps an aim to produce cost-effective optical signal processing devices that can integrate with the present electronics on chip-scale dimensions, as well as can meet the near future demand of more than 1 terabit/s rate. This demand of much increased bandwidth will certainly be unbearable by copper links.

4.2 Energy Band of Silicon

At the side of rich advantages of silicon, the major restriction in using silicon as a light source is related to its indirect band gap structure. This structure implies that the radiative recombination efficiency is low because of the requirement of support of a phonon in order to accomplish the momentum conservation. This actually means that the radiative lifetimes of electron–hole (e–h) pairs have very long that lies in the range of milliseconds. The problem comes from the fact that inside silicon, e–h pairs travel freely having an average distance before recombination is in micrometers.

The probability of undergoing defects or luminescence killer centers therefore is high, even in electronic grade silicon. Consequently, silicon has a non-radiative recombination lifetime of few nanoseconds long, i.e., mostly the recombination of the excited e–h pairs occurs non-radiatively. This leads the internal efficiency to very low level at room temperature. Moreover, high excitation is needed when population inversion is required in order to attain lasing. Under this condition, processes of non-radiativity turn on (such as Auger recombination) at a quicker pace (Fig. 4.1 shows green arrows that represent participation of three particles in non-radiative processes) or free carrier absorption (shown by arrows in orange color in Fig. 4.1). Both of these processes which reduce the excited population and loss mechanisms are also provided. This is the reason why silicon was not considered in the list of light emitter candidates.

In Fig. 4.1, the different arrows are used to indicate the recombination paths for an excited electron and absorption processes. Indirect absorption is indicated by black arrows, whereas indirect radiative recombination with the assistance of a phonon is shown by red arrows. Blue arrows are representing non-radiative recombination, whereas green and orange arrows show Auger recombination and free-carrier absorption, respectively.

Fig. 4.1 Schematic diagram of silicon energy band

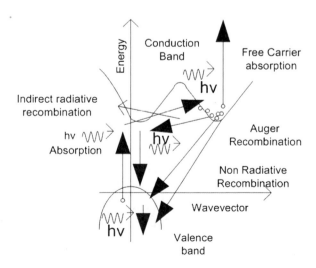

4.3 Types of SOI Waveguides Used in Photonics

Numerous types of silicon waveguides with different structures and light guiding techniques have been presented by various researchers and scientist. These
waveguide structures are strictly application-oriented and manufactured considering assorted parameters like signal wavelength of interest, refractive index contrast (Δn), bending radius, optical absorption and scattering losses, coupling to
the input and output fiber, loss, and so on. Some of the basic types of waveguides
which are not application-oriented are introduced as under:

The four basic types of SOI waveguides are rib, channel, slot (vertical and horizontal), and photonic crystal (PhCs) waveguides. The most common waveguides
being used for silicon photonics are rib and channel waveguides which are shown
in Fig. 4.2b, d. In rib waveguides, guiding layer consists of a slab with a dielectric
rib on top embedded between two low-index-of-refraction layers. The dielectric
rib creates an effective-high-index region just below the rib, with a slightly higher
effective index than the index of the slab, while in channel waveguides the guiding
layer is completely surrounded by a cladding layer. In both the channel and the
rib waveguides, light is confined due to total internal reflection between the high
and low index-of-refraction region [1]. Photonic crystal waveguides, Fig. 4.2c, are
the dielectric structures which are periodic using alternating regions of high and
low refraction indices. In case, the period is of the wavelength scale, high reflec-
tivity is produced with forming of complete photonic bandgap crystal. Silicon is
the suitable material to create single dimensional (1D), two dimensional (2D), and
three dimensional (3D) photonic crystal (PhCs) waveguides [2]. The latter can be
engineered to yield devices with unique properties, such as negative-refraction

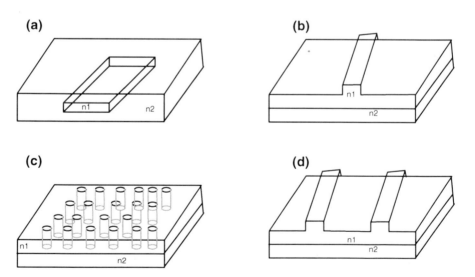

Fig. 4.2 Four configurations of waveguides in silicon. **a** Channel waveguide. **b** Rib waveguide. **c**
Photonic-crystal waveguide. **d** Slot waveguide [1]

lenses, super prisms, self-collimated waveguides, sharp waveguide bends, all-optical buffer memories, dynamic dispersion compensators, and nanoscale 3-D point-defect resonators that provide high Q and high concentration of the optical fields. The uniqueness adds value to a photonic integrated circuit (PIC). In an experimental PIC, conventional Si strip (or rib) waveguides have been joined smoothly to the "unconventional" PhC line-defect waveguides [3]. The slot waveguides, Fig. 4.2d, are based on low index submicrometer slot (air or SiO_2) embedded between two silicon waveguides, this high index contrast produces modes with strong field intensity [4] examples of such waveguides are shown in Fig. 4.3a, b [5].

4.4 Optical Devices for Coupling of Light

Optical fibers and plane SOI waveguides are built for light signal propagation with minimum loss. The planer SOI waveguide devices are mostly constructed with thin layers of different materials of different refractive indices. The optical

Fig. 4.3 General structures of (**a**) vertical and (**b**) horizontal slot waveguides, the light propagates in x direction [1]

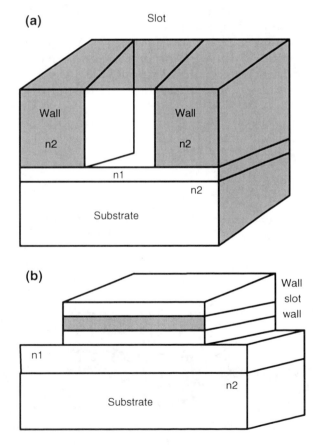

fibers are in the form of strand of glass with circular cross sectioned core at the center surrounded by concentric cladding glass, the fiber core with higher refractive index as compared to cladding maintains the light to propagate inside the core by total internal refraction ($\Delta n \approx 1.45 - 1.40 \approx 0.05$).

As the size of the core (micrometers) of a typical optical fiber is much larger than the core size of SOI planer waveguide (nanometers), the coupling of light from fiber to waveguide may introduce large loss at the interface. The waveguide grating couplers and adiabatic tapers are designed with polished surfaces for coupling of light from/to optical fibers and planer waveguides. There exist a number of techniques to couple light into optical waveguides and/or to collect light from optical waveguides, e.g., butt coupling, grating couplers, tapered couplers, and the evanescent coupling. Each of these techniques has drawback of very high coupling loss, when coupling scheme is applied to wafer scale testing.

Different researchers and manufacturers present their designs for coupling of light with emphasis on the coupling losses, e.g., Tai Tsuchizawa and Koji Yamada et al. came with "the inverse taper approach." Their work shows an efficient fiber-waveguide coupling (a coupling loss of 0.5 dB per connection was obtained), although the modal mismatch is too high between the silicon wire waveguide with effective area ($A_{\text{eff}} \approx 0.1~\mu\text{m}^2$) and a single-mode fiber ($A_{\text{eff}} \approx 50~\mu\text{m}^2$) [6].

S. McNab et al. used the technique of gradual expansion of a core guided mode into a much larger cladding guided mode, which resulted in coupling loss as low as 0.2 dB from a single-mode fiber to a silicon wire waveguide [7, 8]. Another approach, shown in Fig. 3.4 [9, 10] used surface gratings etched onto silicon in which the coupling loss of 1 dB is obtained (Fig. 4.4).

Xia Chen, Chao Li, and Hon Ki Tsang patented their design of optical devices for coupling of light. This design comprises a planar substrate and an optical wave-guiding layer disposed on the planer substrate. The optical wave-guiding layer comprises a grating portion for coupling light between a planar waveguide

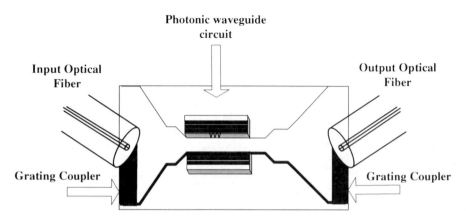

Fig. 4.4 Schematic structures of efficient fiber to waveguide coupling via surface gratings [26]

and an optical fiber; and a tapered guiding portion for converting mode size between the fiber and the planner waveguide. They claim that this design may be implemented as the grating portion which consists of a second grating section having:

1. Uniform periods,
2. Linearly chirped periods, or
3. Optical fiber, which may be positioned normal to a surface of the grating portion [11].

A similar coupler is shown in Fig. (4.5).

The authors [12] introduced a waveguide coupling probe for sending/receiving light into/from an optical waveguide on a substrate, this investigation comprises:

1. An optical element for guiding the light in the propagation direction, and
2. A facet where the light enters or exits the optical element.

The waveguide coupling probe is provided for being placed with the propagation direction under a predetermined angle orthogonal to the waveguide. In [13], Zhaolin Lu and Dennis W. Prather reported an optical coupler for parallel coupling from a single-mode optical fiber or fiber ribbon, into SOI waveguide for integration with silicon optoelectronic circuits. The optical coupler incorporates the advantages of the vertically tapered waveguides and prism couplers, yet offers the flexibility of planar integration. Intel Corporation has recently introduced a method and apparatus for efficient coupling between a silicon photonic chip and an optical fiber [14].

4.5 Nonlinear Optics

The response of any dielectric to light that becomes nonlinear for strong electromagnetic fields, and optical fibers are no exemption. Basically, the origin of nonlinear response is related to inharmonic motion of bound electrons in the influence of an applied field. As a result, the total polarization **P** induced by electric dipoles is not linear in the electric field **E**, but suit the more general relation,

$$\mathbf{P} = \varepsilon_0 \left(\chi^{(1)} \cdot \mathbf{E} + \chi^{(2)} : \mathbf{EE} + \chi^{(3)} \vdots \mathbf{EEE} + \cdots \right) \qquad (4.1)$$

Fig. 4.5 Sample structures of optical coupling device [14]

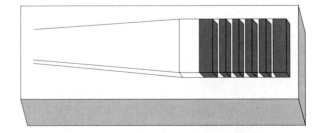

where ε_0 is the vacuum permittivity and $\chi^{(j)}$ is jth order susceptibility. In general, $\chi^{(j)}$ is a tensor of rank $j + 1$. The linear susceptibility $\chi^{(1)}$ represents the dominant contribution to $\mathbf{P}$. Its effects are incorporated through the refractive index n and the attenuation coefficient α.

$$\alpha_{\mathrm{dB}} = -\frac{10}{L} \log \left(\frac{\mathbf{P}_T}{\mathbf{P}_o} \right) \tag{4.2}$$

The second-order susceptibility $\chi^{(2)}$ is responsible for such nonlinear effects as second-harmonic generation and sum-frequency generation. However, it is nonzero only for media that lack inversion symmetry at the molecular level. As SiO_2 is a symmetric molecule, $\chi^{(2)}$ vanishes for silica glasses. As a result, optical fibers do not normally exhibit second-order nonlinear effects (Fig. 4.6).

In nonlinear optics (NLO) we come to know that how light behaves at very high intensities, and also how to take advantage of this behavior to gain remarkable control over light. Basically, nonlinearity is very important phenomenon because it allows light to interact with the light. In a system where electric polarization is only proportional to the electric field, the waveforms that can be superposed exclusive of any change to the dynamics of the individual components. If two waveforms interact, they will cleanly pass through one another. Consequently, if a light is passed with a specified set of frequency in the course of a linear medium, then we can only ever get those frequencies exposed. When the optical nonlinearity is present, conversely the waveforms can mix or self-interact to construct a completely new waveform. Examples of nonlinear phenomena consist of:

(a) In second-harmonic generation where the photons are fundamentally combined to give photons with twice of the energy [15].
(b) For four wave mixing in which three frequency components interact to each other and then construct or amplify a fourth wave [16, 17].
(c) The light with a very broad frequency spectrum that can be generated from a narrow spectrum pulse in super continuum generation [18, 19].

Fig. 4.6 Optical pulse propagation [15]

(d) By introducing the transparency in electromagnetically stimulate in which absorption at a particular frequency is reduced by optically inducing the destructive interference between the parallel quantum states [20].

(e) In Stimulated Raman scattering [21] where the photons are exchanged for photons with shifted energy by means of the attendant excitation or relaxation of the medium; and the modulation instability [22] in which small divergence from a waveform are reinforced by nonlinearity, the foundation is to break up into a chain of pulses.

(f) For exotic linear phenomena the nonlinearity can also be used to create an environment. These include "slow light", a process in which light can be considerably slowed or even halted. The process "fast light" in which the group velocity of light that exceeds the speed of light in a vacuum. It even becomes more negative that is causing a pulse to move toward the source [23].

4.6 Nonlinear Optics

NLO is the branch of optics that explains the behavior of light in nonlinear media; a medium in which the dielectric polarization 'P' acts in response the nonlinearly to the electric field 'E' of the light. It is observed that the nonlinearity exhibits characteristically at very high light intensities such as those provided by pulsed lasers.

4.6.1 Frequency Mixing Processes

(a) Second-harmonic generation abbreviated as SHG or the process of frequency doubling: In general, the light produced in this process is of double frequency which means that the wavelength reduces to half.

(b) Sum frequency generation (SFG) is a process where the frequency of the generated light equals to the sum of two individual frequencies.

(c) Third harmonic generation (THG) is a process in which the frequency of generated light is a tripled frequency. This process is usually completed in two steps: First step of SHG is followed by second step SFG of original and frequency-doubled waves.

(d) Difference frequency generation (DFG) is the process in which light is purely generated having frequency equals to the difference of two other frequencies.

(e) Parametric amplification is fundamentally the signal's amplification of a signal in which the input is present in occurrence of a higher-frequency pump wave, and at the same time the generating of an idler wave that can be considered as DFG.

(f) Optical rectification is basically the generation of electric fields that are quasi-static.
(g) Four-wave mixing (FWM) can also be the result of other nonlinearities.

4.6.2 Linear Versus Nonlinear optics

Linear optics can also be referred as the optics of weak light. It is described as the light is deflected or delayed without its frequency being unaffected.

NLO is also referred as the optics of intense light; it can be described as the special effects that the light itself induces and it propagates through the medium.

4.6.3 Other Nonlinear Processes

(a) Cross-phase modulation (XPM)
(b) Four-wave mixing (FWM)
(c) Cross-polarized wave generation (XPW)
(d) Optical Kerr effect
(e) Intensity dependent refractive index
(f) Self-focusing
(g) Kerr-lens modelocking (KLM)
(h) Self-phase modulation (SPM)
(i) Optical solitons
(j) Raman amplification
(k) Optical phase conjugation
(l) Two-photon absorption
(m) Optical phase conjugation
(n) Brillouin scattering

4.7 Silica on Silicon

The waveguides with step index, having symmetric rectangular cross-section are formed. A layer of SiO_2 wraps the silicon substrate and at core the doped SiO_2 layer is deposited. Photolithography is used to pattern the core and then another layer of SiO_2 is deposited. Figure 4.7 shows the processing steps. The gray layer as shown in figure is of silicon substrate and SiO_2 is deposited on it. On top of this layer, doped SiO_2 shown in dark blue color is deposited, and then lithography and etching are used to pattern this layer. As a last step the conformal deposition of

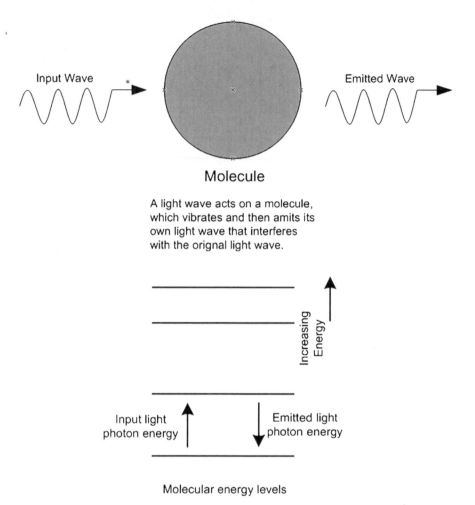

Molecule

A light wave acts on a molecule, which vibrates and then amits its own light wave that interferes with the orignal light wave.

Molecular energy levels

Fig. 4.7 Schematic diagram showing the propagation of light through medium [15]

another layer of SiO_2 is done. The high refractive index core of SiO_2 is yielded in this process on top of Si and over the cladding there is un-doped SiO_2 (Fig. 4.8).

4.8 Silicon in Photonics

The use of silicon in industry has long been established for infrared optics, such as simple lenses, windows, and long-wave detection. Presently, silicon is the material of variety for visible detection, for the most part as the imaging element in digital cameras.

Fig. 4.8 Schematic of fabrication of silica on silicon waveguides [24]

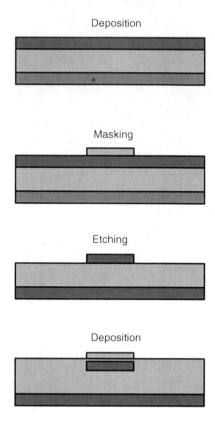

There is uncertainty about the economic and technical advantages of silicon. Generally, it was predictable that silicon would be employed wherever optic fiber is deployed. As estimated nowadays, with the increased demand of internet and data transmission-rate, the need for higher speed, broader bands, and lower cost can effectively be exploited by the under-mentioned material benefits of the silicon:

1. Photonic: provides the transparency of wide band infrared,
2. Electronic: it provides low noise, high speed integrated circuits,
3. Thermal: a good conductor of high heat,
4. Structural: rigged 3-dimensional platforms and packages.

These material properties make possible a wide range of integrated electronic and photonic circuits. Evaluation of the silicon potential can be able to be consulted in the current article by Lipson [25].

4.9 Silicon Photonics Progress in Last Decade

Photonics is becoming progressively more essential in the field of electronics, the reason being that it keeps itself up-to-date with both the "more-Moore", in which the performance is increased when integration and parallelism is increased, and "beyond-Moore", which has new computation principles, advancement trends of electronics. Silicon photonics is the branch which was initiated by Soref in the 1980s [26, 27], is a technology that is capable of combining both photonics and electronics within a single chip in order to get the advantage of both technologies: In this way, the capability of electronics to perform high computation is combined with advantage of high communication bandwidth of photonics. The major attraction of silicon photonics is joined with the opportunity of addition of new functionalities in electronic components, such as wavelength multiplexing, low propagation losses, immunity to electromagnetic noise, and high bandwidth. The key strength of this technology is the silicon properties of non-toxicity, low cost, and very fine ultra-large-scale integrated aka ULSI circuit fabrication technology, which has the responsibility of the huge success of silicon in the field of electronics, can be made most useful. Silicon photonics is the field that is very much promising in the area of research and with the presence of the first commercial devices which are very useful in wide range of applications in reality [28].

Silicon is actually an excellent optical material but suffers from the disadvantage of poor light emitter, the novelty of emission of light from silicon that is porous in nature, at room temperature in 1990 [29] gave major push to the research on light sources based on silicon. Just after that time, the emergence of silicon microphotonics or optoelectronics concept occurred after one another suddenly [30, 31]. The concept of small-sized Silicon-based waveguides was taking its pace; ranges from more than 100 μm^2 usual waveguides that are based on the refractive index contrast provided by distinct doping levels in the duration of years 1980s to 5 μm^2 sizes of rib waveguides, where index contrast was provided by Si/SiO_2.

Material like silicon is very ubiquitous in electronic components and other micro-nanostructures made by man. Increasing significance of academia and industry has caused advanced processing in silicon processing techniques. These processing techniques caused utilization of integrated electronic circuit technology on a vast scale to produce application specific integrated circuits (ASIC), which are playing most significant part in the electronic industry these days. ASICs are not the only products of accurate silicon processing techniques and its advantages—also the advantages continue to plethora demonstrations of micro-opto-electro-mechanical structures (MOEMS) and micro-electro-mechanical structures (MEMS). These microstructures are capable of adding real time sensing to the ICs, resulting in the fabrication of some devices like lab-on-chip commercially for sensing in meteorology and medicine, etc. Currently, some new disciplines like Bio-MEMS, NEMS also known as nano-electro-mechanical structures, etc., are being investigated, aiming to diversifying applications, and growing the

fabrication processes. Silicon-based micro-nano device technology is anticipated
to get more maturity in its status. It may be appropriate to phrase this in the evolution of "lithic" era as the "silicon age."

The nanotechnology's range has been developed as an outcome of continuous
investor's faith consequential from a positive feedback provided by improvements
in the fabrication techniques and machine's limitations in market like cost, size, port-
ability, functionality, etc. Improved processing techniques cause advancement in
devices and characterization methods, have escorted to better consideration of sci-
ence at the nano-scale thus most developed simulation techniques have discovered.

The challenge actually is the development of silicon IC which is compatible
with micro-photonic technology. Moreover, if silicon is the selected material then
merging of these optical devices with CMOS circuits, simplifies the system further
relatively. Silicon photonics may be defined as design and fabrication of photonic

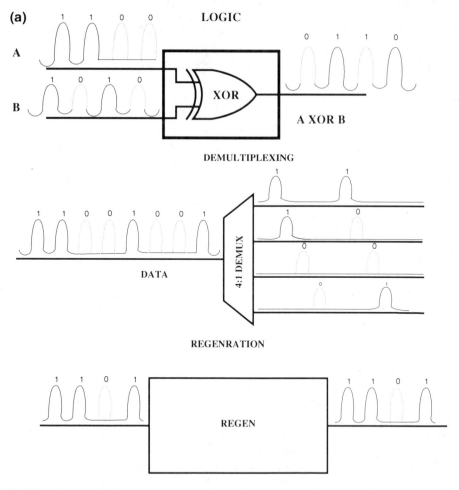

Fig. 4.9 **a** Basic principle/configurations of optical signal processing devices [15]. **b** Basic principle/
configurations of optical signal processing devices [15]

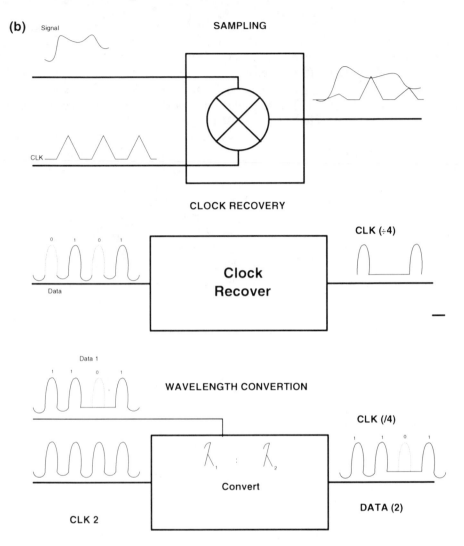

Fig. 4.9 (continued)

devices using conventional Si-CMOS techniques plus it provides substantial cost, size, power savings, and the integration of optics technology to the current electronic-only circuits (Fig. 4.9).

4.10 Lifespan of Silicon Photonics

Silicon photonics underwent a significant boom for the last couple of decades. The spectacular efforts have been invested in this field. Many vital breakthroughs have taken place on light emitters [32–35], waveguides [36–39], modulators [40, 41], resonators [42–44], and detectors [45]. Silicon photonics also catch the attention of the interest of industry. Many of companies and industries are eager to perform research and get real commercial opportunities [46, 47]. ST-Microelectronics described the highly efficient electroluminescence (EL) from an Er-doped device in year 2002 [48]. In 2003, the photonic band gap waveguides with very low losses were presented by IBM [49]. Low-loss silicon wire waveguides and a 30 GHz SiGe photo detector were established by IBM in 2004 [50].

In electronic industry a modulator with modulation bandwidth exceeding 1 GHz was fabricated by Intel [51]. In addition, wavelength conversion and all-optical switching in silicon were projected [52–54]. By Intel in year 2005, a continuous wavelength (CW) silicon Raman laser was introduced [55]. A 10 Gbps modulator was demonstrated independently both by Intel and Luxtera [56, 57]. A hybrid silicon evanescent laser was invented by the University of California Santa Barbara and supported by Intel in 2006 [58]. Also by Cornell a broadband amplifier based on Raman gain was introduced [59]. Moreover, the electro-optical effect in strained silicon was demonstrated in 2006 [60].

In 2007, at Intel the device performance reached up to 40 Gbps for active silicon photonic devices. The devices are a mode-locked silicon evanescent laser, and a fast Ge photo-detector and also modulator [61, 62]. Luxtera launched its first photo receiver that is a four-channel 10 Gbps monolithic optical receiver used in 130 nm CMOS with integrated Ge waveguide photo detector [63]. The optical buffering of 10 bits at 20 Gbps in 100 cascaded ring resonators [64] and recently, the fast optical switching [65] demonstrated by IBM team. The recent development given an idea about that silicon photonics area is on the mount as it involves the invention of new structures and application of new materials or of new phenomena in accessible materials.

Silicon nanocrystal (Si-nc, Si-ncs) embedded in a dielectric matrix. In most cases the silicon oxide is one of the important materials, which has already made great contributions to the breakthroughs which were pointed out above. It will also continue to improve the performance of various kinds of these devices (Fig. 4.10).

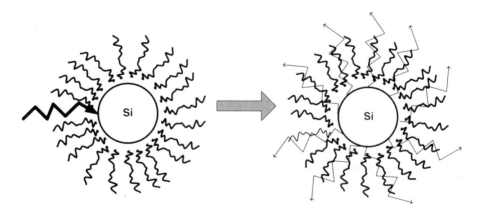

Fig. 4.10 Silicon nano-crystal as an emitter, which emits UV-light after excitation by a laser (http://ocmp.phys.rug.nl/Research/Nanoscale%20dynamics/index.html)

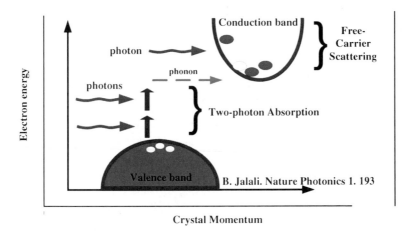

Fig. 4.11 Two photon and free carrier absorption [67]

In present days, one of the limitations in electronic circuits is the interconnection of technologies. The most up-to-date IC chips technology has practically touched 109 on-chip transistors by means of using 4 nm thick gate oxides and 100 nm minimum feature size running at 4 GHz clock speeds [66]. It is believable that in materials and fabrication technology the incremental advances in chip architecture will compliment improvements. However, signal propagation delay (between transistors) is still larger than device gate delay (delay within device) (Fig. 4.11).

4.11 Two-Photon Absorption

Two-photon absorption or TPA is defined as the absorption done simultaneously between two photons that can be of the same frequency or unlike frequencies, for exciting a molecule from one state (usually the ground state) to another state, i.e., electronic state of higher energy. The difference between the energy involving lower states and upper states of the molecule is equivalent to the summation of the individual energies of the two photons.

Two-photon absorption is numerous times of magnitude weaker than the linear absorption. It also differs from linear absorption in a way that the strength of absorption which depends on the square of the light intensity, hence it is also termed as nonlinear optical process.

References

1. Lipson M (2005) Guiding, modulating, and emitting light on silicon—challenges and opportunities. Lightwave Technol 23:4222–4238
2. Joannopoulos JD, Meade RD, Winn JN, Johnson SG (2008) Photonic crystals: molding the flow of light: one dimensional photonic crystals, 2nd edn. pp 44–104

3. Reed GT, Mashanovich GZ, Headley WR et al (2006) Issues associated with polarization independence in silicon photonics. Quantum Electron 12:1335–1344
4. Almeida VR, Xu Q, Barrios CA, Lipson M (2004) Guiding and confining light in void nano-structure. Opt Letts 29:1209–1211
5. Xie M, Yuan Z, Qian B, Pavesi L (2009) Silicon nanocrystals to enable silicon photonics. Opt Letts 7:319–324
6. Tsuchizawa T, Yamada K, Fukuda H et al (2005) Microphotonic devices based on silicon microfabrication technology. Quantum Electron 11:232–240
7. McNab S, Moll N, Vlasov Y (2003) Ultra-low loss photonic integrated circuit with mem-brane-type photonic crystal waveguides. Opt Express 11:2927–2939
8. Roelkens G, Dumon P, Bogaerts W, Thourhout DV, Baets R (2005) Efficient fiber to SOI photonic wire coupler fabricated using standard CMOS technology. In: LEOS 18th Annual Meeting, Sydney, Australia, Oct 2005
9. Taillaert D, Bogaerts W, Bienstman P et al (2002) An out of plane grating coupler for effi-cient butt-coupling between compact planar waveguides and single-mode fibers. Quantum Electron 38:949–955
10. Narasimha A (2004) Low dispersion, high spectral efficiency, RF photonic transmission sys-tems and low loss grating couplers for silicon-on-insulator nanophotonic integrated circuits. PhD dissertation, University of California Los Angeles, 2004
11. Chen X, li C, Tsang HK (2009) US20090290837
12. Scheerlinck, Thourhout V, Beats R (2009) WO2009003969
13. Lu Z, Prather DW (2008) US007428358
14. Liu A (2010) US7643710
15. Tomasi W (2001) Advanced Electronic communication system, 5th edn, Prentice Hall
16. Inoue K (1992) Four-wave mixing in an optical fiber in the zero-dispersion wavelength region. J Lightwave Technol 10:1553–1561
17. Slusher RE, Hollberg LW, Yurke B, Mertz JC, Valley JF (1985) Observation of squeezed states generated by four-wave mixing in an optical cavity. Phys Rev Lett 55:2409–2412
18. Ranka JK, Windeler RS, Stentz AJ (2000) Visible continuum generation in air-silica micro-structure optical fibers with anomalous dispersion at 800 nm. Opt Lett 25:25–27
19. Dudley JM, Genty G, Coen S (2006) Supercontinuum generation in photonic crystal fiber. Rev Mod Phys 78:1135–1184
20. Fleischhauer M, Imamoglu A, Marangos JP (2005) Electromagnetically induced transpar-ency: optics in coherent media. Rev Mod Phys 77:633–673
21. Gisela Eckhardt RW, Hellwarth FJ, McClung SE, Schwarz DW, Woodbury EJ (1962) Stimulated Raman scattering from organic liquids. Phys Rev Lett 9:455–457
22. Agrawal GP (2001) Nonlinear fiber optics. Academic Press, London
23. Boyd RW, Gauthier DJ (2002) Slow and fast light. Prog Opt 43:497–530
24. Lockwood D, Pavesi L (2004) Silicon photonics II: topics in applied physics, vol 119. Springer, Berlin
25. LipsonML (2005) J Lightwave Technol. Guiding, Modulating and Emitting Light on Silicon-Challenges and Opportunities 23:4222–4238
26. Soref RA, Lorenzo JP (1985) Single-crystal silicon-A new material for 1.3 and 1.6 μm inte-grated-optical components. Electron Lett 21:953–954
27. Soref RA, Lorenzo JP (1986) All-silicon active and passive guided-wave components for $\lambda = 1:3$ and $1:6$ μm. IEEE J Quant Electron 22:873–879
28. Pavesi L, Lockwood D (2004) Silicon photonics: topics in applied physics, vol 94. Springer, Berlin
29. Canham LT (1990) Silicon quantum wire array fabrication by electrochemical and chemical dissolution of wafers. Appl Phys Lett 57:1046–1048
30. Soref RA (1993) Silicon-based optoelectronic. Proc IEEE 81:1687–1706
31. Bisi O, Campisano SU, Pavesi L, Priolo F (1999) Silicon based microphotonics: from basics to applications. In: Proceedings of E. Fermi Schools: course CXLI, Amsterdam, The Netherlands

32. Pavesi L, Dal Negro L, Mazzoleni C, Franzo G, Priolo F (2000) Optical gain in Si nanocrystals. Nature 408:440–444
33. Nayfeh MH, Barry N, Therrien J, Akcakir O, Gratton E, Belomoin G (2001) Stimulated blue emission in reconstituted films of ultrasmall silicon nanoparticles. Appl Phys Lett 78:1131–1133
34. Boyraz O, Jalali B (2004) Demonstration of a silicon Raman laser. Opt Express 12:5269–5273
35. Chen M, Yen J, Li J, Chang J, Tsai S, Tsai C (2004) Stimulated emission in a nanostructured silicon pn junction diode using current injection. Appl Phys Lett 84:2163–2165
36. Lee KK, Lim DR, Luan H-C, Agarwal A, Foresi J, Kimerling LC (2000) Effect of size and roughness on light transmission in a Si/SiO_2 waveguide: Experiments and model. Appl Phys Lett 77:1617–1619
37. Loncar M, Doll T, Vuckovic J, Scherer A (2000) Design and fabrication of silicon photonic crystal optical waveguides. J Lightwave Technol 18:1402–1411
38. Han H-S, Seo S-Y, Shin JH (2001) Optical gain at 1.54 μm in erbium-doped silicon nanocluster sensitized waveguide. J Appl Phys 27:4568–4570
39. Vlasov YA, O'Boyle M, Hamann HF, McNab SJ (2005) Active control of slow light on a chip with photonic crystal waveguides. Nature 438:65–69
40. Png CE, Reed GT, Atta RMH, Ensell GJ, Evans AGR (2003) Development of small silicon modulators in silicon-on-insulator (SOI). Proc SPIE 4997:190–197
41. Kuo Y-H, Lee Y-K, Ge Y, Ren S, Roth JE, Kamins TI, Miller DAB, Harris JS (2005) Strong quantum-confined stark effect in germanium quantum-well structures on silicon. Nature 437:1334–1336
42. Akahane Y, Asano T, Song BS, Noda S (2003) High-Q photonic nanocavity in a two-dimensional photonic crystal. Nature 425:944–947
43. Xu Q, Schmidt B, Pradhan S, Lipson M (2005) Micrometre-scale silicon electro-optic modulator. Nature 435:325–327
44. Song BS, Noda S, Asano T, Akahane Y (2005) Ultra-high-Q photonic double-heterostructure nanocavity. Nat Mater 4:207–210
45. Michel J, Liu JF, Giziewicz W, Pan D, Wada K, Cannon DD, Jongthammanurak S, Danielson DT, Kimerling LC, Chen J, Ilday FO, Kartner FX, Yasaitis J (2005) High performance Ge p-i-n photodetectors on Si. In: Proceedings of group IV photon conference, pp 177–179
46. Jalali B (2006) Silicon photonics. J Lightwave Technol 24(12):4600–4615
47. Soref RA (2006) The past, present, and future of silicon photonics. IEEE J Sel Topics Quantum Electron 12:1678
48. Castagna ME, Coffa S, Monaco M, Muscara A, Caristia L, Lorenti S, Messina A (2003) High efficiency light emitting devices in silicon. Mater Sci Eng B 83:105
49. McNab S, Moll N, Vlasov Y (2003) Ultra-low loss photonic integrated circuit with membrane-type photonic crystal waveguides. Opt Express 11:2927–2939
50. Koester SJ, Schaub JD, Dehlinger G, Chu JO, Ouyang QC, Grill A (2004) High-efficiency, Ge-on-SOI lateral PIN photodiodes with 29 GHz bandwidth. In: Proceedings of device research conference
51. Liu A, Jones R, Liao L, Samara Rubio D, Rubin D, Cohen O, Nicolaescu R, Paniccia M (2004) A high-speed silicon optical modulator based on a metal-oxide-semiconductor capacitor. Nature 427:615–618
52. Raghunathan V, Claps R, Dimitropoulos D, Jalali B (2004) Wavelength conversion in silicon using Raman induced four-wave mixing. Appl Phys Lett 85:34–36
53. Almeida VR, Barrios CA, Panepucci RR, Lipson M (2004) All optical control of light on a silicon chip. Nature 431:1081–1084
54. Boyraz O, Koonath P, Raghunathan V, Jalali B (2004) All optical switching and continuum generation in silicon waveguides. Opt Express 12:4094–4102
55. Rong H, Liu A, Jones R, Cohen O, Hak D, Nicolasecu R, Fang A, Paniccia M (2005) An all-silicon Raman laser. Nature 435:292–294
56. Liao L, Samara-Rubio D, Morse M, Liu A, Hodge D, Rubin D, Keil UD, Franck T (2005) High speed silicon Mach-Zehnder. Opt Express 13:3129–3135

57. Gunn C (2006) CMOS photonics for high-speed interconnects. IEEE Micro 26:58–66
58. Fang AW, Park H, Cohen O, Jones R, Paniccia M, Bowers J (2006) Electrically pumped hybrid AlGaInAs-silicon evanescent laser. Opt Express 14:9203–9210
59. Foster MA, Turner AC, Sharping JE, Schmidt BS, Lipson M, Gaeta AL (2006) Broadband optical parametric gain on a silicon photonic chip. Nature 441:960–963
60. Fage-Pedersen J, Frandsen LA, Lavrinenko A, Borel PI (2006) A linear electrooptic effect in silicon, induced by use of strain. In Part of: 2006 EEE/LEOS international conference on proceedings of 3rd group IV photon, pp 37–39
61. Yin T, Cohen R, Morse M, Sarid G, Chetrit Y, Rubin D, Paniccia MJ (2007) 31 GHz Ge n-i-p waveguide photodetectors on silicon-on-insulator substrate. Opt Express 15:13 965–13 971
62. Liu A, Liao L, Rubin D, Basak J, Nguyen H, Chetrit Y, Cohen R, Izhaky N, Paniccia M (2007) Silicon optical modulator for high-speed applications. In: Proceedings of 4th IEEE international conference group IV photon, pp 1–3
63. Masini G, Capellini G, Witzens J, Gunn C (2007) A four-channel, 10 Gbps monolithic optical receiver in 130 nm CMOS with integrated Ge waveguide photodetectors presented at the National fiber optic engineers conference, 2007, Paper PDP31, unpublished
64. Xia F, Sekaric L, Vlasov Y (2007) Ultracompact optical buffers on a silicon chip. Nat Photon 1:65–71
65. Vlasov Y, Green WMJ, Xia F (2008) High-throughput silicon nanophotonic deflection switch for on-chip optical networks. Nat Photon 2:242–246
66. Kimberling LC Devices for silicon microphotonic interconnection
67. Jalali B (2006) Silicon photonics. J Lightwave Technol 24(12):4600–4615

Chapter 5
High-Nonlinearity in Glass Fibers and Cross-Phase Modulation

Abstract In this chapter, the effects of nonlinearity in optical fiber communication link have been discussed and highlighted. Desirable effects of nonlinearity in general and undesirable effects, in particular, in an optical fiber are enumerated with examples. Cross-phase modulation (XPM) and high-nonlinearity (HNL) glasses along with their advantages and disadvantages are spelled out with the help of their theories and applications in practical systems.

Keywords Nonlinear • Pulse broadening • Polarization fluctuation • Pulse walk-off

5.1 Introduction to Linearity and Nonlinearity

Nonlinearity is very interesting and important phenomena particularly in the field of engineering and its applied areas. Usually, a misconception prevails about nonlinearity and its subsequent applications in engineering systems, which in fact, is incorrect. On the contrary, nonlinear behavior can effectively be exploited for many useful applications in engineering design and analysis. In order to understand nonlinearity in true prospective, the under-mentioned examples from daily life could be helpful for the purpose.

While driving a vehicle, when a driver presses accelerator, the car speeds up proportionally to the acceleration, if the car gains 100 km/h speed in 6 s, this is a total linear behavior.

On the other hand, if a person visits a departmental store where a single bar of chocolate costs \$5; however, in case a deal is offered, that is, if a person buys more than one bar of chocolate, 20 % discount is offered. This exhibits a nonlinear behavior as the purchased items (chocolates) is not proportional to the money paid. Both of the above-mentioned examples complement the fact that nonlinearity can be at times useful as well.

J. Ahmed et al., *Optical Signal Processing by Silicon Photonics*, SpringerBriefs in Materials, 53
DOI: 10.1007/978-981-4560-11-5_5, © The Author(s) 2013

Now let us be more specific, in electronic systems, normally nonlinearity is undesirable. This phenomenon can be seen if we observe the characteristic curve of a silicon diode. There exists a nonlinear region that is up to 0.7 V and once the barrier potential exceeds, we enter the linear region.

The most important property of linearity is nothing but the validity of principle of superposition, as a consequence of this attribute; a linear system performs satisfactorily when excited by a standard test signal. Also the amplitude of the test signal is unimportant since any change in input signal amplitude results simply in change of scale in output having no change in basic characteristics output.

Conversely, the response of a nonlinear system to a particular test signal is no guide to their behavior to other inputs because the principle of superposition no longer holds. In fact the non linear system response may be highly sensitive to input amplitude.

5.2 Nonlinearity in Physical Systems

Primarily, there are two common types of nonlinearity i.e., *incidental nonlinearity* and *intentional nonlinearity*. The former nonlinearity is the type which is already present in the system. In these systems, it is strived to design a system in such a way that the adverse effect of nonlinearity is restricted. Some of the common incidental nonlinearity includes saturation, dead zone, coulomb friction, backlash, etc.

On the other hand, the nonlinearity which is deliberately inserted in the system with an intention to modify the system's characteristics is called intentional nonlinearity. The most common example is the operation in relay.

5.3 Nonlinearities in Optical Fibers

Subsequent to the understating and basic concepts of linearity and nonlinearity in physical systems, the same can comprehensibly be transformed to communication systems which exploit optical fiber as a main source of medium.

The study of nonlinearity in optical fibers was started somewhere in 1970s and no remarkable progress was made in this area up to 1980. As a result, the study of nonlinearity was not given due attention and thus ignored till 1980s. A considerable amount of work was done during 1980s and then onward in 1990s, the subject was drew an attention of the concerned quarters particularly in optical fiber-based communication. Since then, an extraordinary amount of research has been taken up and is being carried out in this specific area. It is believed that nonlinearity may be taken over in this decade that will certainly be a major breakthrough in optical communication systems.

5.4 Major Nonlinear Effects

As already discussed, nonlinearity plays an important role in the output response of the systems. Following are the major nonlinear effects encountered in optical fiber communication systems.

(a) Stimulated Brillouin Scattering (SBS)
(b) Stimulated Raman Scattering (SRS)
(c) Cross-Phase Modulation (XPM)
(d) Self-Phase Modulation (SPM)
(e) Four-Wave Mixing (FWM).

The role of nonlinear fiber devices has been highly appreciated in the recent years. This appreciation has been given to nonlinear fiber devices due to their built-in property of ultrafast response time along with the potential application in optical communication systems. In order to generate sufficient nonlinear phase shifts, long fibers are required because nonlinearities of conventional silica-core silica-clad fibers are too low. Nonetheless, these long length devices cause under-mentioned serious problems which have to be traded-off one way or other.

(a) Pulse walk-off
(b) Pulse broadening
(c) Polarization fluctuation.

The above-mentioned problems act as limiting factors for response time, bandwidth, and maximum attainable bit-rate. Thus, in view of the prevailing scenario, the shorter length is preferably important in order to achieve ultrafast switching and higher bit-rate data transmission. Same time, we simultaneously need shorter length and higher nonlinearities to achieve our goals so in order to increase the fiber nonlinearities, selection of high-nonlinearity (HNL) materials is of prime importance too. These materials include Litharge, Bismite, Tellurite, and Chalcogenide glasses which serve the purpose of nonlinearity as such. Along with high nonlinearity, these materials also exhibit high-group velocity dispersion and high losses. These factors adversely affect the performance of nonlinear fiber devices. Therefore, it has to be taken into account too that how HNL glasses affect the performance of nonlinear fiber devices, so both the advantages and the disadvantages has to be considered in parallel fashion. For this purpose, we can evaluate different types of fibers which are constructed from different types of HNL glasses. By using HNL glasses for fibers, we can effectively reduce the device length of nonlinear fiber devices. In addition to the length, other problems of group velocity dispersion (GVD) and losses could also be controlled because of the extremely short device length. Once the above-mentioned effects are under control, walk-off, pulse broadening, and polarization fluctuation in nonlinear fiber devices can also be suppressed. Thus, this would lead to ultrafast switching and higher bit-rate. In this respect, a scheme of wavelength division demultiplexing is in place that is based upon the optical Kerr effects. The most important attribute of this scheme is that the adverse effects of GVD of HNL glasses can be changed into an advantage of wavelength division demultiplexing. Resultantly, it would yield increased maximum transmittable bit-rate in optical communication by simultaneously demultiplexing optical time division multiplexed signals and wavelength division—multiplexed signals with an optical Kerr effect-based demultiplexer.

In order to evaluate different fibers which are constructed from high nonlinear glasses, we must meticulously understand the problems that need to addressed and rectified on priority. As mentioned previously, these problems are related to device length, optical power, the nonlinear-index coefficient, effective mode area and nonlinear phase shift, the pulse walk-off caused by group velocity mismatch, the pulse broadening caused by GVD, and the polarization fluctuation caused by environmental disturbances.

5.4.1 Pulse Walk-Off

As discussed in the previous section, the important feature of GVD is that pulses at different wavelengths propagate at different speed due to the group velocity differences. It causes pulse walk-off that plays a central role in nonlinear fiber switching devices involving two or more closely spaced pulses. In nonlinear fiber switching devices, they generally use different wavelengths for control and signal pulses to separate them easily after switching. Therefore, the effect of pulse walk-off between control and signal pulses seriously affects the switching performance. The interaction between the two pulses terminates when the fast-moving pulse completely walks through the slower moving pulse. More specifically, when walk-off exceeds one time slot of optical time division multiplexing (OTDM) signal pulses, the control pulse interacts with many other different signal pulses, making it hard to select only one channel of the signals. Finally, it limits the bit-rate of the signal and switching bandwidth. This feature is governed by the walk-off parameter d_w given by:

$$d_w = \left| \beta_1 \left(\lambda_1 \right) - \beta_1 \left(\lambda_2 \right) \right| = \left| \frac{1}{v_g \left(\lambda_1 \right)} - \frac{1}{v_g \left(\lambda_2 \right)} \right|$$

The walk-off length for the initial pulse width of τ can be defined as

$$L_W = \frac{\tilde{L}}{d_w}$$

As the walk-off parameter increases and the initial pulse width decreases, the walk-off length decreases, thus making it hard to achieve high bit-rate and broad switching bandwidth.

5.4.2 Pulse Broadening

Another issue in nonlinear fiber devices is the temporal broadening of the control pulse caused by GVD. Even though the control and signal pulses have the same group velocities, if the control pulse has high GVD, the temporal broadening of the control pulse reduces the switching time of devices. A shorter control pulse width exhibits faster response time, but it is often limited due to GVD. The dispersion length LD is already defined and the critical pulse width, τ, is defined as: [1]

$$[\tau_c]^2 = 4\ln(2\,|\beta_2|\,L)$$

where β_2 is the dispersion parameter and L is the fiber length. Over a propagating length L, an optical pulse broadens by a factor of 2 when $L = LD$. Therefore, to avoid the pulse broadening effect, the following condition should be satisfied.

$$\frac{L}{L_D} \ll 1$$

and the output pulse width τ_0 given by

$$[\tau_o]^2 = [\tau_i]^2 \left[1 + \left(\frac{\tau_c}{i}\right)^2\right]$$

should be nearly equal to the input pulse width τ_i.

Usually, nonlinear fiber devices require long device-lengths to induce a sufficient nonlinear phase shift at lower optical powers due to the low nonlinearities of silica fibers. Furthermore, since the minimum switching time is limited by the control pulse width; a short control pulse is required to achieve much faster switching. These requirements of long device length, which increases the walk-off parameter and GVD, and short control pulse in nonlinear fiber devices make the switching performance much worse because a short pulse is broadened even more for a given fiber length. Therefore, the device length should be short enough to minimize the walk-off parameter and pulse broadening due to GVD. This can be achieved by introducing HNL glasses since they can increase the nonlinear-index coefficient and decrease the effective mode area, thus resulting in extremely short devices.

5.4.3 Polarization Sensitivity

The fluctuation of SOP severely degrades the performance of nonlinear fiber devices because SOP is very critical to the operation of almost all nonlinear fiber devices. Therefore, polarization controllers or PMFs are usually used in order to control the SOP of an optical pulse, since the SOP in standard fibers changes randomly as the pulse propagates along the fiber. However, it is not efficient to adjust the polarization controllers continuously whenever required, and moreover, huge built-in stress birefringence in conventional low-index cladding PMFs depends highly on temperature variation. In particular, nonlinear fiber devices based on the optical Kerr effect always use PMFs to ensure that the control pulse which is linearly polarized in either one of the two principal axis maintains its SOP along the fiber, because the optical Kerr effect operates based on the variation of SOP. In such devices, since the direction of the linearly polarized signal pulse makes a 45° angle with the two principal axes, the signal pulse will suffer PMD by the built-in birefringence. Moreover, this intentional nonvarying built-in birefringence in PMFs is much higher than the varying birefringence in standard fibers. As a result, the use of PMFs also brings serious problems which cause the limitation of response time and bit-rate of signals. The problems include the signal pulse broadening caused by PMD and the pulse walk-off between the control and signal pulses caused by the birefringence-induced group velocity mismatch. For the non-linear fiber devices based on mode interference, SOPs of the two modes should be maintained and remain the same during operation for better performance, but it is hard to achieve due to temperature dependence of the SO. Short lengths of devices are able to solve these problems of the signal pulse broadening caused by PMD and the pulse walk-off by the birefringence-induced group velocity mismatch. Devices having short lengths are hardly affected by environmental disturbances. Therefore, the use of HNL glasses for short devices would be the better way to effectively reduce polarization sensitivity.

5.5 Cross-Phase Modulation

Many nonlinear dispersive systems like optical fiber communications exhibit instability, known as the modulation instability, which has already been discovered in Physics [2–10]. In case, weak perturbations grow exponentially causing interplay between nonlinearity and GVD then this process is known as modulation instability. But as far as optical fiber is concerned, modulation instability occurs due to anomalous dispersion, so, manifests itself as breakup of the continuous waves/quasi continuous waves radiation in a stream of extremely short pulses [6–10]. Anomalous dispersion is also necessary for solitons [11, 12] which is the result of a balance between the nonlinearity induced SPM and GVD. However, recently it has been observed that the observation of modulation instability with cw propagating beams is hampered by competing nonlinear effects like SBS [9, 10], under quasi cw conditions. These experiments were performed in infrared region beyond 1.3 μm (micrometer), in order to operate in anomalous dispersion regime of silica fiber.

When two or more optical fields propagate inside the fiber then a new type of modulation instability occurs, even in the normal dispersion regime. But the actual mechanism behind this phenomenon is XPM which occurs due to nonlinear phase change of an optical field induced by other co-propagating fields. The modulation instability induced by XPM has fundamental importance as it suggests the possibility of the soliton formation in the normal regime and it also has practical implications for the propagation of visible radiation in optical fibbers.

5.5.1 Theory

In order to present the major results in a possible simplest way, let us consider the case of two optical fields propagating in a single mode fiber. The field amplitudes are $A1$ and $A2$, respectively that satisfy the nonlinear Schrodinger equation [11, 12] modified for XPM by the addition of cross coupling term:

$$i\left(\frac{\partial Aj}{\partial z} + V_{gj}^{-1}\frac{\partial Aj}{\partial t} + \frac{1}{2}\alpha_j A_j\right) = \frac{1}{2}\beta_j\frac{\partial^2 Aj}{\partial t^2} - \gamma_j\left(|A_j|^2 + 2|A_{3-j}|^2\right)A_j \quad (5.1)$$

where $j = 1$ or 2, V_g shows the group velocity, α_j is the absorption coefficient ($\beta_j = dvg^{-1}/d\omega$ and $\beta_j < 0$ for anomalous dispersion) and

$$\gamma_j = n_2\frac{\omega_j}{C A_{eff}} \quad (5.2)$$

related to fiber nonlinearity that is responsible for the both SPM and XPM. In Eq. (5.2), A_{eff} denotes the effective core area for silica fibers. And the last term in Eq. (5.1) is because of XPM that couples the two waves. It is the XPM-induced coupling due to that modulation instability in the normal dispersion regime occurs provided that $\beta_j > 0$ for both waves. For the sake of simplification, we neglect the fiber loss by setting $\alpha_j = 0$ because inclusion of the fiber losses do not change the basic concept of the subject. Steady-state solution of Eq. (5.1) is given in the following equation showing that

$$\overline{A_j} = \sqrt{P_j}e^{(i\phi_j)}, \quad j = 1, 2 \quad (5.3)$$

P_j relates to optical power and phase

$$\phi_j = \gamma_j\left(P_j + 2P_{3-j}\right)Z \quad (5.4)$$

We can assume the stability of the steady state by letting

$$A_j = \left(\sqrt{P_j} + a_j\right)e^{(i\phi_j)} \quad (5.5)$$

Showing α_j as weak perturbation, linearizing equation (5.1) in a_1 and a_2 gives

$$i\left(\frac{\partial a_j}{\partial z} + V_{gj}^{-1}\frac{\partial a_j}{\partial t}\right)$$
$$= \frac{1}{2}\beta_j\frac{\partial^2 a_j}{\partial t^2} - \gamma_j P_j\left(a_j + a_j^*\right) - 2\gamma_j\left(P_1 P_2\right)^{\frac{1}{2}}\left(a_{3-j} + a^*_{3-j}\right) \quad (5.6)$$

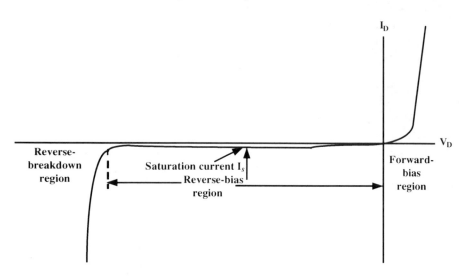

Fig. 5.1 Characteristic curve of a silicon diode

where $j = 1, 2$. If we suppose a general solution of the form

$$a_j = u_j \cos(K_z - \Omega t) + i V_j \sin(K_z - \Omega t) \tag{5.7}$$

where K is wave number and Ω is the modulation frequency, Eq. (5.6) gives a set of homogeneous equations containing u_1, u_2, v_1 and v_2. If K and Ω satisfy the dispersion relation then this set of equations will have a nontrivial solution (Fig. 5.1)

$$\left[(K - \Omega/V_{g1})^2 - f_1 \right] \left[(K - \Omega/V_{g2})^2 - f_2 \right] = C^2 \tag{5.8}$$

where,

$$f_j = \frac{1}{2} \beta_j \Omega^2 \left(\frac{1}{2} \beta_j \Omega^2 + 2\gamma_j P_j \right) \tag{5.9}$$

and then parameter of coupling i.e., C is given by

$$C = 2\Omega^2 (\beta_1 \beta_2 \gamma_1 \gamma_2 P_1 P_2)^{\frac{1}{2}} \tag{5.10}$$

$$(K - \Omega/V_{g1})^2 = \frac{1}{2} \left\{ (f_1 + f_2) \pm \left[(f_1 + f_2)^2 + 4 \left(C^2 - f_1 f_2 \right) \right]^{\frac{1}{2}} \right\} \tag{5.11}$$

Clearly, K will become complex if $C^2 > f_1 f_2$ is satisfied. Solving Eqs. (5.9) and (5.10), the condition becomes

$$\Omega^2 < \Omega_c^2 = \frac{1}{2} \left\{ \left[(b_1 + b_2)^2 + 12 b_1 b_2 \right]^{\frac{1}{2}} - (b_1 + b_2) \right\} \tag{5.12}$$

Fig. 5.2 Gain spectra $g(\Omega)$ for different power levels P_1 and P_2 of the two beams [9]

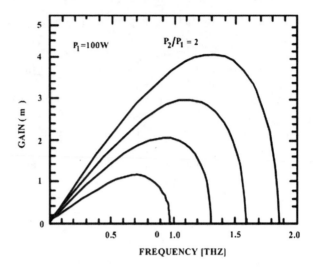

where

$$b_j = \left(\frac{4\gamma_j}{\beta_j}\right) P_j \tag{5.13}$$

So for frequencies such that $|\Omega| < \Omega_c$ then a weak modulation of the steady state will be affected by the gain given below

$$g\left(\Omega\right) = 2\mathrm{Im}\left(K\right) \tag{5.14}$$

where Im is used for the imaginary part.

Figure 5.2 shows the gain spectra against different values of the ratio of power i.e., P_2/P_1 by taking $P_1 = 100$ W. As the gain spectrum is symmetric on both sides $g(-\Omega) = g(\Omega)$ so only one part is shown.

The parameters of fiber related to a realistic situation of propagation of silica fibers in visible region near about 0.53 µm. The wavelengths of two propagating fields differ very slightly so that β_j, γ_j, and V_{gj} are nearly the same for $j = 1$ and 2. In this respect, g can be evaluated as a closed form and can be expressed as

$$g\left(\Omega\right) = \beta_1 \left|\Omega\right| \left(\Omega_c^2 - \Omega^2\right)^{\frac{1}{2}} \tag{5.15}$$

Where Ω_c is given in Eq. (5.12). The maximum gain occurs at $\Omega = \Omega_c/\sqrt{2}$ and the maximum gain increases as we increase P_1 and P_2. Since the gain is because of XPM and it vanishes when either the powers P_1 or P_2 become zero means both the optical fields should be present.

If K has an imaginary part against different values of the Ω, then the steady state becomes unstable as a_1 and a_2 will exponentially grow with the fiber length. This phenomenon shows the modulation instability [2–10] because it leads to modulation of steady-state amplitude. But in the absence of XPM, $C = 0$, then Eq. (5.8) has the solution $k = \Omega / V_{gj} \pm \sqrt{f_j}$ for $j = 1, 2$. So this is the dispersion relation as obtained previously [6–10] for single field. Since, $\beta_j > 0$ for the normal dispersion regime so $f_j > 0$ from Eq. (5.9), it then shows that K is real and the instability of modulation does not occur for the normal dispersion regime.

But the situation becomes totally different when XPM couples the two propagating optical fields. As Eq. (5.8) shows that K is becoming complex for some values of Ω, although both β_1 and β_2 are positive. It is observed that if we neglect the group velocity mismatch and consider $V_{g1} = V_{g2}$ for the time being then from Eq. (5.8), it shows that modulation instability to be occur.

From practical point of view, it is mandatory to present group velocity mismatch by the parameter

$$\delta = \left| v_{g1}^{-1} - V_{g2}^{-1} \right| \qquad (5.16)$$

Figure 5.2 is showing the gain spectra for different values of δ obtained by using the Eqs. (5.8) and (5.14), when both powers are same ($P_1 = P_2 = 100$ W). Other parameters are almost identical to the values those used for Fig. 5.2. As δ increases and the gain spectrum becomes narrow and it also shifts toward higher frequencies having peak height approached about 6 1/m. But the main point to be observed is that group velocity mismatch is not detrimental to the existing instability of modulation.

$$V_m \cong \delta / 2\pi\beta_1 \qquad (5.17)$$

As peak gain is not sensitive to group velocity mismatch for the case $\delta > 2$ ps/m, that is shown in Fig. 5.3 where the peak gain is plotted as a function of power ratio P_2/P_1 against different values of δ and $P_1 = 100$ W.

The results shown in Figs. 5.2, 5.3 and 5.4 show that XPM-induced modulation instability are observable under real experimental situations. For instance the case of two beams propagating in visible region at about 0.53 μm with

Fig. 5.3 Effect of group velocity mismatch δ on gain spectra for equal power case i.e., $P_1 = P_2 = 100$ W. Other parameters are same in Fig. 5.2 [9]

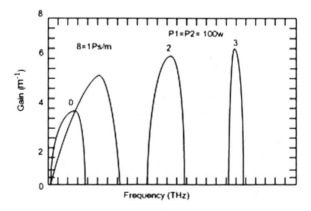

Fig. 5.4 Peak gain variation versus the ratio of power i.e., P_2/P_1 for different GVM [13]

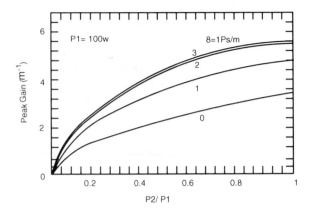

wavelengths a few nanometers separated. The parameter $\delta = 2\text{--}5$ ps/m for a wavelength in the range 5–10 nm, the peak gain nearly equal to 6 1/m can be expected for approximately equal powers of 100 W as shown in the Fig. 5.4.

Optical spectra of both the beams that are propagating in response to modulation instability have to be divided into side-bands present with a frequency separation on both sides as given in Eq. (5.17). When $\delta = 2$ ps/m and $\beta = 0.06$ ps^2/m, the frequency is about Vm = 5 THz which is nearly equal to 5 nm at 0.53 μm. The side-bands produced as a result of modulation are because of input noise amplification provided by vacuum fluctuations or spontaneous emission with a peak gain of about 6 1/m, the amplification factor of exp(6L) for length L meter optical fiber. Hence, a few build-ups of side-bands are expected for a fiber having few meters in length. As far as time–domain response is concerned, both the propagating beams perform amplitude modulation with period 1/Vm in femto-second range which is 200 fs in the example mentioned above.

Therefore, unstable modulation manifest itself as breaking each beam of cw into a stream of extremely short pulses with repetition rate of some tera hertz. Taking the scenario of conventional modulation that is not stable [9, 10] experimental observation of modulation instability produced by XPM, requiring the optical pulses to stop SBS. So the overlap of both pulses will occur throughout the fiber length if $\tau >> \delta L$.

So far the assumption has been made that both the propagating beams are happening, respectively, over the optical fiber but there is another important option that if the other beam is produced as a result of SRS. In case of silica fibers, stoke shift is of 13 THz i.e., (at $\lambda = 0.53$ μm, 12 nm) and the mismatched group velocity that lies between the stoke waves and pump is δ nearly equal to 5 ps/m. However, the analysis described here is not relevant and can not be applied directly because if the effect of Raman gain is not included, possibly the qualitative feature of XPM based modulation instability is there and from Eq. (5.17), modulation frequency is also about 14 m i.e., nearly 13 THz for a 0.53 μm pump. Hence modulation side-bands will appear at a number of Stoke shift thereby making it difficult identifying existence of Stoke lines of higher order.

There is another possibility that if we consider that the interaction of the two spectral components which are mostly dominant of Single-Phase Modulation and the spectrum of single pulse are broadened. Two side-bands appearing in the pulse spectrum at the same time are showing that a connection is presented in this analysis. It is observed from these experiments that the normal dispersion regime side-bands can be interpreted as XPM-induced modulation instability [13, 14].

5.6 High-Nonlinearity Glasses

Optical glasses with fast response time and high third-order nonlinearities are very promising materials for ultrafast nonlinear fiber devices [15, 16]. The third-order optical nonlinearity is the most important property for realization of ultrafast all-optical switching. This section addresses the general properties of HNL Glasses of Lead-Oxide (PbO, Litharge), Bismuth-Oxide (Bi_2O_3, Bismite), Tellurium-Oxide (TeO_2,Tellurite), and Chalcogenide glasses with a focus on their nonlinear properties such as the third-order susceptibility and nonlinear-index coefficient. The characteristics of the conventional and microstructure fibers made of those HNL glasses are also discussed.

5.6.1 Bismite (Bi_2O_3) Glasses

Bismite is derived from the ignition of bismuth nitrate which in turn is obtained from the heavy metal bismuth. The latter is very similar to lead; however, there is no evidence of toxicity and it has been used in low-temperature frits, as a flux in conductive glazes and in metal enamels. Bismite has been used instead of litharge in amounts up to 50 % in optical glasses to improve durability and to increase the specific gravity and the refractive index. Arsenic is often used with it to prevent a tendency toward grey coloration. Bismite is a very effective substitute for litharge, providing the same high gloss, flow, bubble clearance characteristics, high refractive index, surface tension, viscosity, and resistance to aggressive dish washer detergents. Bismite melts at a lower temperature than litharge and thus glazes can be even more fluid. The properties of Bi_2O_3-based glasses are summarized below:

(a) High linear refractive index (1.87 ~ 2.6)
(b) High nonlinear-index coefficient ($32 ~ 1,810 \times 10 - 20[m^2/W]$)
(c) Very wide transmission range (0.45 ~ 5 μm)
(d) Transition temperature (~500 °C) and melting temperature (~900 °C)
(e) Fusion-spliceable to SiO_2-based glasses
(f) Easy integration to silica-based systems
(g) High mechanical, chemical, and thermal durability
(h) Simple and easy fabrication process
(i) No toxicity.

5.6.2 Tellurite (TeO₂) Glasses

Tellurite glasses, compared to silicate and fluoride glasses, have a reasonably wide transmission region, the lowest phonon energy among the common oxide glasses, larger refractive indices, and high nonlinear-index coefficients. Their large refractive index and small phonon energy are desirable for radioactive transitions of rare-earth ions (Er^{3+}, Tm^{3+}, Nd^{3+}, Pr^{3+}, Yb^{3+}, and so on) and the application of fiber lasers and amplifiers. The lower photon energy leads to lower nonradioactive transition rate (high fluorescence quantum efficiency) between adjacent rare-earth energy levels, causing new fluorescence transitions, and laser emission from additional energy levels. Under normal conditions, tellurium-dioxide (TeO_2) has no vitrification ability without modifiers. Thus, glass-modifiers and/or secondary glass-formers are necessary in order to obtain tellurite glasses. Incorporation of a second component to tellurite glasses is expected to extend the Te–O inter atomic distance, which should increase the mobility of polyhedron thereby provide a favorable condition for tellurite vitrification. The properties of TeO_2-based glasses are summarized below:

(a) High linear refractive index (1.82 ~ 2.27)
(b) High nonlinear-index coefficient ($16 \sim 210 \times 10 - 20[m^2/W]$)
(c) Very wide transmission range (0.35 ~ 6 μm)
(d) Low transition temperature (250 ~ 400 °C) and melting temperature (450 ~ 800 °C)
(e) Good glass stability, strength, and corrosion resistance
(f) Good rare-earth ion solubility
(g) Relatively low phonon energy for oxide glasses ($600 \sim 850$ cm^{-1}) to minimize nonradioactive losses
(h) High electrical conductivity
(i) High resistance to devitrification and atmospheric moisture
(j) Good chemical durability
(k) High homogeneity
(l) Highly capable of incorporating large concentrations of rare-earth ions into the matrix.

5.6.3 Chalcogenide Glasses

The chalcogenide glasses are composed of one or more chalcogenide elements such as S, Se, and Te, with other elements such as As, Ga, Ge, In, and Sb to form a stable glasses. Other elements such as P, I, Cl, Br, Cd, Br, Ba, Si, or Tl can be added to these glasses for tailoring their thermal, mechanical, and optical properties. The chalcogenide elements are covalent in nature and can exhibit chain, ring, and/or network structures. In contrast to oxide glasses, they can depart from atomic stoichiometry through the partial segregation of chalcogens and/or red ox adjustments of the non-chalcogen constituents. Chalcogenide glasses are melted in closed systems due to the chalcogens volatility. Therefore, it is typically required to seal the batch components in an evacuated silica ampoule and to increase temperature slowly with rocking motion to promote mixing. The glass components are often individually prepurified to remove oxide and hydride impurities which can impair the mid-IR transmission. The most distinct characteristics of chalcogenide glasses are high refractive indices and ZDWs in the mid-IR range. It has been confirmed that Se-based chalcogenide glasses possess higher nonlinearity than S-based chalcogenide glasses, and it seems that Te-based chalcogenide glasses are not suitable for optical communications at wavelengths of 1.3 or 1.5 μm because their band gap wavelengths become longer than the operating wavelengths.

Besides, the chalcogenide glasses based on Ga–La–S are of great interest, particularly due to their low toxicity, high transition temperature, and excellent rare-earth solubility. The chalcogenide glasses based on As–S are suitable candidates for passive and active fiber applications due to their crystallization stability and their mechanical and chemical stability. The properties of chalcogenide glasses are summarized below:

(a) High linear refractive index (2.03 ~ 3.23)
(b) High nonlinear-index coefficient (101 ~ 9000 × 10−20[m^2/W])
(c) Ultra wide transparency region from near IR to far IR (0.7 ~ 16 μm)
(d) Low transition temperature (150 ~ 250 °C) and melting temperature (300 ~ 400 °C)
(e) High chemical durability
(f) Low phonon energy (200 ~ 300 cm^{-1})
(g) Wide glass-forming region
(h) Can be fabricated into low-loss fiber
(i) Can be doped by rare-earth elements.

5.6.4 Advantages of High-Nonlinearity Glasses

HNL glasses such as Litharge, Bismite, Tellurite, and Chalcogenide have both high linear refractive indices and high-nonlinear-index coefficients. Due to their high nonlinear-index coefficients, nonlinear fiber devices made of high-nonlinear glasses can have highly nonlinear properties. In addition, high linear refractive indices of the HNL glasses enable higher index difference between the core and cladding of the devices, thus making it possible to develop smaller effective mode area. This also leads to increase of nonlinearities in the nonlinear fiber devices. Furthermore, if the HNL glasses are used in microstructure fibers, the effective mode areas can be reduced further due to much higher index difference and thus, the nonlinearities of the devices can be increased much more. As a result, it is certain that one can increase the nonlinear properties of nonlinear fiber devices by introducing the HNL glasses with high linear refractive index and high nonlinear-index coefficients. Since nonlinearities of silica glasses are very small, conventional nonlinear fiber devices usually require either long length or high power for a sufficient nonlinear phase shift in order to achieve the desired nonlinear effects. The long length of fiber leads to the serious problems of pulse walk-off, pulse broadening, and polarization fluctuation. It is also certain that one can drastically reduce the required device length by introducing the HNL glasses due to their high nonlinearities and thus can suppress the problems caused by the long length. Moreover, the fabrication processes of HNL glasses are easier because of their low melting temperatures and high glass stability. Also, they have wide transmission range from near-infrared to far-infrared.

5.6.5 Disadvantages of High-Nonlinearity Glasses

However, the HNL glasses have extremely high losses and GVDs. This extremely high-group velocity dispersion can cause the deleterious effects of high relative group delay and rapid pulse walk-off between two pulses at different wavelengths, despite the short device length, due to extremely high-group velocity difference. It also can cause huge pulse broadening and high loss limits the performance of the devices. These problems can be major limiting factors for response time, switching bandwidth, and maximum transmittable bit-rate.

References

1. Slusher RE, Lenz G, Hodelin J, Sanghera J, Shaw LB, Aggarwal ID (2004) Large Raman gain and nonlinear phase shifts in high-purity As2Se3 chalcogenide fibers. J Opt Soc Am B 21(6):1146–1155
2. Benjamin TJ, Feir JE (1967) J Fluid Mech 27:417
3. Ostrovskiy LA (1966) Zh Eksp Teor Fiz 51:1189 [Sov Phys JETP 24:797 (1967)]
4. Karpman VI (1967) Pis'maZh EkspTeor Fiz 6:829 [JETP Lett 6:277 (1967)]
5. Hasegawa A (1975) Plasma instabilities and nonlinear effects. Springer, Heidelberg
6. Hasegawa A, Brinkman WF (1980) Tunable coherent IR and FIR sources utilizing modulational instability. IEEE J Quantum Electron 16:694
7. Hasegawa A (1984) Generation of a train of soliton pulses by induced modulational instability in optical fibers. Opt Lett 9:288
8. Andersonand D, Lisak M (1984) Modulational instability of coherent optical-fiber transmission signals. Opt Lett 9:468
9. Tai K, Hasegawa A, Tomita A (1986) 1100 × optical fiber pulse compression using grating pair and soliton effect at 1.319 μm. Phys Rev Lett 56:135
10. Trillo S, Wabnitz S (1991) Dynamics of the nonlinear modulational instability in optical fibers. Opt Lett 16(13):986–988
11. Hasegawa A, Tappert F (1973) Transmission of stationary nonlinear optical pulses in dispersive dielectric fibers. I. Anomalous dispersion. Appl Phys Lett 23:142–144
12. Mollenauer LF, Stolen RH, Gordon JP (1980) Experimental observation of picosecond pulse narrowing and solitons in optical fibers. Phys Rev Lett 45:1095–1098
13. Tomlinson WJ, Stolen RH, Johnson AM (1985) Optical wave breaking of pulses in nonlinear optical fibers 10(9):457–459

14. Valk B, Vihelmsson K, Salour MM (1987) Appl Phys Lett 50:656
15. Lines ME (1991) Oxide glasses for fast photonic switching: a comparative study. J Appl Phys 69(10):6876–6884
16. Friberg SR, Smith PW (1987) Nonlinear optical glasses for ultrafast optical switches. IEEE J Quantum Electron 23(12):2089–2094

Chapter 6
Supercontinuum Generation by Nonlinear Optics

Abstract Supercontinuum (SC) sources are replacement of white light sources. Supercontinuum generation in optical fiber, pumped by different sources, which include pumping with femto second (fs), picoseconds (ps) pulse sources, and continuous sources are reviewed in this chapter. The nonlinear Schrödinger equation is used to discuss the spectral broadening or SC generation. Recently, SC generation has been proved very effective and some of its applications are very promising for future ultra-high bandwidth networks.

Keywords Supercontinuum • Spectral broadening • Nonlinear Schrödinger equation • White light • Pulse compression • Spectroscopy • Optical coherence tomography

Abbreviations

fs   Femto second
ps   Picoseconds
SC   Supercontinuum
FWM   Four-wave mixing
SOI   Silicon-on-insulator
XPM   Cross-phase modulation

6.1 Introduction

Nonlinear optics has inherent feature of generation of new frequency components and spectral broadening, the process is known as supercontinuum generation. This phenomenon occurs when narrowband incident pulses from pump beam results in spectral broadening due to nonlinear effects. This gives a broadband, very often

J. Ahmed et al., *Optical Signal Processing by Silicon Photonics*, SpringerBriefs in Materials,   69
DOI: 10.1007/978-981-4560-11-5_6, © The Author(s) 2013

a white light continuous output [1]. The observation of second harmonic generation in 1961 was soon followed with the innovation of a large number of nonlinear processes [2], including stimulated Raman scattering (SRS), stimulated Brillouin scattering (SBS), Self-phase modulation (SPM), the Kerr effect, and four-wave mixing (FWM). Nonlinear effects in optical fibers have become an area of academic research and gained great importance in the optical fiber-based systems [2].

SC generation was first reported in 1970 by Alfano and Shapiro, after propagation of picoseconds pulses with 5 mJ pulse energy centered at 530 nm in bulk BK7 (borosilicate fibers) glass. They investigated a white light spectrum which covering the range from 400 to 700 nm is generated [3]. In the late 1990, the advent of photonic crystal fiber (PCF), a new class of optical waveguides, gained widespread attention throughout the scientific community and paved way to a revolution by generating ultra-broadband high brightness spectra through SC generation [1]. The characteristics of PCFs that led to such practical importance are their guidance properties that result in single-mode propagation over wide wavelength ranges, modal confinement, high nonlinearity, and the ability to give their group velocity dispersion [1].

Spectral broadening was first observed in CS2 by Brewer in 1967. The spectral composition of the self-trapped filaments was observed not to correspond to Raman, Rayleigh, or Brillouin scattering. These experiments were repeated in CS2 by Shimizu who observed a regular periodic structure superimposed on the frequency-broadened spectrum. These results were interpreted in terms of a molecular orientation model. Polloni et al. studied self-focusing and SPM in CS2 and interpreted their results in terms of a librational model in which the molecules "rock" in the field of the neighboring molecules. Brewer and Lee used a mode locked 1.06 μm laser to study self-trapping in low viscosity liquids, high viscosity liquids, and glass where linear and rotational diffusion modes are frozen out. Their results support a molecular electronic distortion mechanism for the index of refraction change [4].

Husakou and Herrmann (2001) give the numerical modeling of SC generation in PCF using femtosecond pulses and the role of soliton fission in the spectral broadening process was highlighted for the first time [5]. There are several applications of SC generation in different fields such as pulse compression, spectroscopy, optical coherence tomography, and the design of tunable ultrafast femtosecond laser sources [2].

Regarding telecommunication systems, fiber-based SC sources have been of much interest due to their stable design, low potential cost, compact package, and high reliability [6]. Research has been done for SC generation in fiber lasers and in optical fibers pumped by different light sources which include fs and ps pulse sources and continuous-wave (CW) light sources.

6.2 Supercontinuum Generation

Alfano and Shapiro, after having their first observation about SC generation, reported the white light spectrum generation in bulk BK7 glass which covered the complete visible range that lies between 400 and 700 nm, when 5 mJ ps pulses were propagated at 530 nm [3]. They carried out this work for identifying the no resonant FWM or four-photon coupling for the first time. This experiment concluded that the generated lights spectral width is 10 or more times wider than anything investigated previously. The authors do not highlight this characteristic in their publication particularly. The phrase "super continuum" was not even used in their paper; sometime later (Manassah et al. 1984) introduced it [7]. Bondarenko et al. (1970) reported results which were not far different from them [8]. Interestingly, at that time, nonlinear spectral broadening of laser light was not completely new, Stoicheff (1963) had already observed it [9].

Spectral broadening is the phenomena that had also been observed in CS2 [4] and interpreted correctly in terms of the SPM nonlinear process. In the meantime, the SC generation phenomena was referred to as anomalous frequency broadening, super broadening [8], or white light continuum.

The SC phenomenon is also reported as wide frequency range white light sources [10]. SC production is a highly complex process in bulk material, involving a complex coupling between temporal and spatial effects. On the other hand, a temporal dynamic process is involved in SC generation in optical fibers, with the characteristics of transverse mode which can only be determined by linear waveguide properties. As a matter of fact, it is suggested that a further inspiration in studying the SC generation in PCFs helps in clarifying the nature of temporal nonlinear propagation effects if improvement is required in understanding of more complex spatiotemporal (bulk) case [1]. Femtosecond pulses of low intensity are used to observe supercontinuum generation [11].

In 2004, a new way was demonstrated by the generation of spectral broadening in highly nonlinear photonic crystal fibers by the contribution of SPM and FWM [12]. Later on in 2005, work was done on sub-wavelength diameter waveguides that reduced the threshold power for generating SC [13].

In 2006, spectral broadening in nonlinear fibers occur over range from 900 to 1,400 nm by using nanosecond noise pulses from Ytterbium-doped fiber amplifier (YDFA) [14]. In 2007, spectral broadening of more than 600 nm was investigated by launching noise like pulses of energy 10 nJ into high nonlinear fiber [15].

In 2010, SC generation was taken up over wavelength range of 1,100–1,800 nm. This is achieved by pumping the pulses with the peak power of 66 W, centered at wavelength of 1,550 nm into arsenic sulfide fiber, uniformity in SC is also reported [16].

6.3 SC Generation in Photonic Crystal Fibers

Usual optical fibers consist of two concentric glass cylinders typically having different refractive indices. In case of the refractive index of the outer cladding lower than the inner core, the assistance takes place through total internal reflection at the boundary of core and cladding. Conventionally, most of the fibers have the core-cladding refractive index difference smaller (0.1 %), and a lot of the propagation characteristics are acquiescent for analysis [17].

The term "photonic crystal fiber", having confined light and it is guided through a photonic band gap (PBG) effect. On the other hand, light guidance can happen in one of the two ways depending on the specific PCF geometry. Conditionally, when the fiber has a hollow core in the center of the structure, actual PBG guidance can take place. PCFs of this type are significantly attractive and interesting just because their advantage for distortion-free and lossless transmission, optical sensing, particle trapping, and for novel nonlinear applications in optics.

This is not that type of PCF of hollow-core that are used in SC generation experiments. In fact, SC generation is reported in PCFs with a solid core that lies in the center of the structure, so that the fiber comprises solid glass region surrounded by an air holes array running along its length. In this scenario, the central region's effective refractive index of the PCF is higher than that of the surrounding air-hole region, and guidance takes place with the help of total internal reflection that is modified [18]. Although this concept is similar to the typical fiber's guidance mechanism, additionally, the degrees of freedom presented by modification in the periodicity and hole size in such an index-guiding PCF open up opportunities available to engineer the fiber waveguide properties in ways that are nonexistent in standard fiber.

The SC generation is observed in PCFs, over a much wider source parameters range than from conventional fibers or bulk media. A bulk material has capability of generating broad spectra comparable to complex sources requirement of with pulse energies at micro joule levels. However, SC generation in PCFs at nano joule levels with the additional advantage of broadband single mode guidance properties of the PCF caused generation of SC while retaining a uniform spatial profile, on the other hand, the case often reported in bulk experiments where SC generation was often related to filamentation effects.

The SC generation in PCFs applications are in many fields such as optical spectroscopy, coherence tomography, and, particularly, in optical frequency metrology.

The factors for determining the generation of SC are the peak power, the dispersion of the fiber relative to the pumping wavelength and the length of the pulse. The dispersion is determined differently for different type of nonlinear effects that are taking part in the continuum formation, and eventually for the spectrum nature in terms of spectral stability and shape.

6.3.1 Photonic Crystal Fibers

The phrase "photonic crystal fiber" is motivated by this fiber class's unique cladding structure. Standard fibers have capability of directing light by total internal reflection between a core having a high refractive index (usually germanium doped silica), surrounded in a cladding having a lower index (usual fluorine-doped or pure silica). The differences of indexes in PCFs are obtained by matrix formation of different material with low or high refractive index. Thus creating a hybrid material with properties that are unobtainable in solid materials (e.g., very low

Fig. 6.1 Classical triangular
cladding single-core photonic
crystal fiber schematic
which shows guided light
in a solid core embedded in
air holes triangular lattice.
The hole-size, d, determine
the structure of fiber and the
hole-pitch, Λ [17]

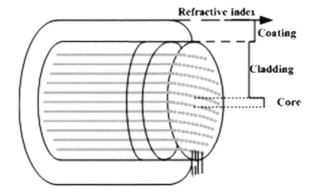

index or novel dispersion), the hybrid material cladding can be constructed with
a similar structure that is found in certain crystal, which is the origin of the term
photonic crystal fiber. The fibers are not fabricated in crystalline materials as the
name might indicate.

There are two fundamental classes of PCFs: Index-guiding PCFs and fibers that
confine light through a PBG (Fig. 6.1).

A PCF that is index-guided, is usually made up of a solid glass high index core
embedded in a cladding having air-filled structure where a number of air holes
are arranged in a way that forms a model running along the fiber length, air-silica
material of hybrid nature is created having lower refractive index even lower than
that of core. The matrix structure of air-silica has helped in creating numerous
other names like holey fibers and micro-structured, but all of them are referred to
same type of fiber even if the terminology is different.

Figure 6.2 shows the single-core classical triangular cladding photonic crystal
fiber's principle, which has been proved as one of the most flexible and efficient
designs, which has emerged as the basis for most of the PCFs today. The type of
fiber in Fig. 6.2 is a fiber of large mode area. The fiber's outer diameter is normally of 125 μm and the fiber's pitch shown in the figure is therefore 10–15 μm
accordingly. The pitch for normally designed nonlinear fibers is of 1–3 μm
approximately, and the nonlinear PCF's micro-structured region therefore, only
takes up the inner 20–50 % of the fiber cross section.

Fig. 6.2 a SEM a multimode
fiber having zero disper-
sion at visible wavelengths.
b Optical microscopic schematic of a zero dispersion
single-mode fiber at wavelengths around 800 nm. The
hole-sizes relatively are ~0.9
and 0.5, respectively [19]

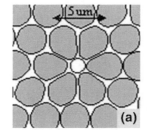
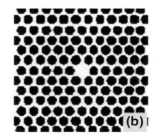

Two basic structures of nonlinear PCFs are: *Multimode fibers* with microstructure like a cobweb and having extremely small core, *Single-mode fibers* with smaller holes, slightly larger cores, and engineered zero dispersion [19].

6.4 Numerical Modeling of SC Generation

In optical fibers spectral broadening of incident pulses or SC generation is governed by nonlinear effects like SPM, SRS, and FWM. Figure 6.3 shows the spectral broadening. In this incident pulse (blue) is launched into photonic crystal fiber and result in SC generation (red line).

The numerical modeling of SC can be done by solving the nonlinear Schrödinger equation (NLS). To include intrapulse Raman scattering and dispersive effect, the following is the equation of pulse evolution inside single mode fibers can be used [20].

$$
\frac{\partial A}{\partial z} + \frac{1}{2}(\alpha(\omega_o) + i\alpha_1 \frac{\partial}{\partial t})A + \beta_1 \frac{\partial A}{\partial t} + \frac{i\beta_2}{2}\frac{\partial^2 A}{\partial t^2} - \frac{\beta_3}{6}\frac{\partial^3 A}{\partial t^3}
$$
$$
= i\left(\gamma(\omega_o) + i\gamma_1\frac{\partial}{\partial t}\right)\left(A(z,t)\int\limits_0^\infty R(t')|A\left(z,t - t'\right)|^2 dt'\right) \tag{6.1}
$$

where amplitude of pulse envelope is presented by $A(z, t)$, β_2 is group velocity dispersion (GVD) parameter is nonlinear parameter responsible for SPM, the integral is representation of energy transfer in intrapulse Raman scattering. For

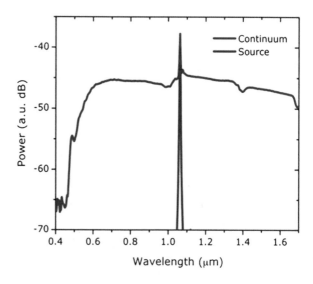

Fig. 6.3 Spectral broadening (*red line*) by launching incident pulse (*blue line*) in photonic crystal fiber [20]

numerical modeling of SC, we can use Eq. (6.1) by adding higher order dispersion terms. Then this will yield the following equation:

$$
\frac{\partial A}{\partial z} + \frac{1}{2}(\alpha(\omega_o) + i\alpha_1 \frac{\partial}{\partial t})A + \sum_{m=2}^{M} i^{m-1}\frac{\beta_m}{m!}\frac{\partial^m A}{\partial t^m}
$$

$$
= i\left(\gamma(\omega_o) + i\gamma_1\frac{\partial}{\partial t}\right)\left(A(z,t)\int_0^{\infty} R\left(t'\right)|A\left(z,t-t'\right)|^2 dt'\right)
\tag{6.2}
$$

where M is order representing the limit of dispersive effect, β_m is coefficient of Taylor's series expansion of propagation constant (t) is the nonlinear response that includes Raman contribution.

$$
\dot{R}(t) = (1 - f_r)\delta(t - t_e) + f_r h_r(t)|A(z,t - t')|^2
$$

where t_e represents short delay which is often negligible, fractional contribution of delayed Raman response, and Raman response function are represented by f_r and $h_r(t)$.

Equation (6.2) gave successful result in modeling some features of SC by using ultra short pulses in nonlinear fibers. The split step Fourier transform method can be used to solve the equation. Here the choice of M is not clear. It can $M = 6$ for numerical simulation or values as large as 12 for some experiments. In fact, to all orders dispersion can be included numerically because in split step method dispersive effect is carried out in spectral domain of pulse by ignoring all nonlinear terms [20].

The term on right-hand side of equation is due to several nonlinear effects like SPM, SRS, and self steepening.

In Fourier domain, Eq. (6.2) can be written as

$$
\sum_{m=2}^{\infty}\frac{\beta_m}{m!}(\omega - \omega_o)^m
$$

$$
= \beta(\omega) - \beta(\omega_o) - \beta_1(\omega_o)(\omega - \omega_o)
\tag{6.3}
$$

where $\beta_1 = 1/v_g$, v_g represents group velocity and is centered frequency [21]. This approach needs knowledge about $\beta(\omega)$ or propagation constant over entire frequency range over SC can spread. In case of tapered fiber or fiber with constant core cladding refractive indices, $\beta(\omega)$ can be obtained by solving the Eigen value equation. This approach is inappropriate for highly nonlinear fibers like photonic crystal fibers [20].

6.5 SC Generation with Picosecond Pulses

In 1993, SC generation acted as an ideal source for WDM optical systems. It was used to generate pulse-train at multiple wavelengths by using picoseconds pulses at 1.55 μm [20]. Research has been done for high average power SC sources; a picoseconds laser generates the pulses with duration of 14 ps at repetition rate of 480 MHz and that exhibit 20 W average output power. This generates the SC over spectral range of 11 nm with average power of 7 W [22]. Ultra short high pulse SC is generated by using a sub-picoseconds erbium fiber laser source that operates at wavelength of 1.557 μm and give peak power more than 1 MW [23]. SC is generated in [24] by using Er-doped fiber chirped laser at 1.55 μm. This generates the pulses with duration of 1 ps and 2 μJ pulse energy at repetition rate of 200 kHz. As the pulse energy is so high it results in generating a smooth SC at wavelength region of 1.3 μm shown in Fig. 6.2 using 1 ps pulses and dispersion-shifted fibers.

6.6 SC Using Continuous Wave Pumping

Picoseconds pulses are not required for generating spectral broadening as ns-pulses gave result of SC generation at high power levels. Even CW pumping lasers led to spectral broadening at high power levels [20]. In Travers et al. [25] the first CW source is investigated that generates SC in visible spectral region 400 W CW pump source with average power of 50 W is used to led spectral power densities of 50 mW/nm and spectral broadening over 1,300 nm. CW Raman fiber laser centered at the wavelength of 1455.3 nm generates SC over more than 200 nm bandwidth [12]. SC is reported in the wide spectral range of 1,200–1,780 nm [26]. This is achieved by using the continuous waves in highly nonlinear fibers (Fig. 6.4).

6.7 SC Generation with Femtosecond Pulses

The use of femtosecond pulses gained practical importance with the advent of highly nonlinear fibers. SC using femtosecond pulses is generated with the power conversion efficiency of 55 % in optical fibers. The spectral characteristic of this SC depends upon power radiation and pulse wavelength of pump. High spectral densities of 300 mW/nm and average power densities of approximately

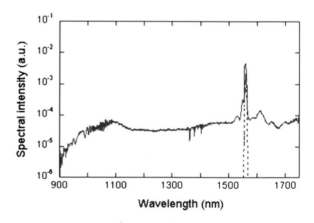

Fig. 6.4 Ultra wideband SC generation using 1 ps pulses and dispersion shifted fibers [24]

0.3 mW/nm for peak power of input radiation achieved within the spectral range of 530–1,100 nm. Such sources with peak powers at nanometer can be used for nanoscale systems [27]. 100 fs pulses at 1,560 nm were launched into sub-5 cm section of highly nonlinear fiber exhibit spectral broadening over 700 nm and third harmonic radiation over 520 nm [28]. 250 fs pulses from Yb:KYW laser with the average power of 150 mW centered at the wavelength of 1,046 nm has been supplied to amplifier input. In this spectral broadening with spectrum width of 50 nm and with average spectral power density of higher than 65 mW/nm is achieved [29].

6.8 Applications of SC Generation

SC sources are the substitute of white light, tungsten-based sources which are used in characterization of broadband attenuation, spectroscopy, and microscopy setups.

The major drawback of white light sources is the low brightness that is finding out by the filament temperature. Larger filaments with same brightness are used in sources with high power at output; therefore same power can be coupled to single mode fiber.

There is also an issue of coupling of all incandescent light sources to fiber. The light coupling efficiency is very low that results in small fraction of light will be available into fiber. The SC sources resolve both the issues of coupling and intensity and give the possibility of making sources that has spectral width like white light sources and intensity of a laser.

There is big problem of pump source in replacement sources of white light with the SC. An incandescent light or white light sources are cost-effective (can be construct from few of hundred dollars) and takes very small space of laboratory. On the other hand, SC sources needs large pump source (a large femtosecond laser system) and this would not build up more compact and cost-effective SC sources. This problem can be resolved by increasing the bandwidth of spectral broadening and reducing power at threshold [16]. This is achieved by using sources of picoseconds or nanosecond pulses centered at 1,060 nm wavelength that leads to less cost, portable, and compact devices [19].

Most SC experiments yield output in the mW-range, but systems with high average power have also been realized. For example, a 900 nm broad SC source with an average output power of 2.4 W was recently demonstrated [20]. The output was generated by pumping a 1 m long PCF with zero dispersion wavelengths at 975 nm. The fiber was pumped by a mode-locked Nd: YVO4 laser with a pulse length of 10 ps, repetition rate of 85 MHz, and an average power of 5 W. The complete system is simple and compact ($500 \times 250 \times 100$ mm^3) and potentially cost-effective.

Therefore, poor resolution is caused by long wavelengths because of the reason that the spectral width of the source is not large satisfactorily. Sources for OCT typically are sources based on amplified spontaneous-emission, super luminescent diodes, and ASE, from semiconductors or doped fibers. For all these sources the common things are restrictions to wavelength range and limited spectral bandwidth. SC sources have contributed well in resolving these issues, where band-width is responsible for astonishing resolution enormously. A longitudinal resolution of only 2.5 μm can be achieved in this way using a 1,210 nm broad SC range lies between 390 and 1,600 nm, produced from a Ti:Sapphire mode-locked laser by using 2 nJ 100 fs pulses [6].

The pumped systems having a range 1,060 nm are supposed to be very useful, as a large flat stable continuum mostly by Raman scattering is created by the slow pulses. The desired wavelength range can be selected by filtering large continuum-filtering which is obtainable easily in, e.g., band gap guiding fibers, in which light is made to be guided in only a limited wavelength range.

References

1. Dudley JM, Genty G, Coen S (2006) Supercontinuum generation in photonic crystal fiber. Phys Rev Lett 24:584–587
2. Agrawal GP (2011) Non linear fiber optics: its history and recent progress [invited]. J Opt Soc Am B Opt Phys 28(12):A1
3. Alfano RR, Shapiro SL (1970) Emission in the region 4000 to 7000 Å via four-photon coupling in glass. Phys Rev Lett 24:592–594
4. US army armament command Faran Lford, Arsenal Philadelphia Pennsylvania 19137 Self phase modulation: a reviews (1975)
5. Husakou AV, Herrmann J (2001) Supercontinuum generation of higher-order solitons by fission in photonic crystal fibers. J Opt Soc Am B 19:2171–2182
6. Li S, Ruffin AB, Kuksenkov DV, Li M-J, Nolan DA (2007) Supercontinuum generation in optical fibers: invited paper. Proc of SPIE 6781:678105
7. Manassah JT, Ho PP, Katz A, Alfano RR (1984) Ultrafast supercontinuum laser source. Photonics Spectra 18:53–59
8. Bondarenko NG, Eremina IV, Talanov VI (1970) Broadening of spectrum in self-focusing of light in crystals. Sov J Exp Theoret Phys Lett 12:85–87
9. Stoicheff BP (1963) Characteristics of stimulated Raman radiation generated by coherent light. Phys Lett 7:186–188
10. Nishizawa N (2009) Octave spanning high quality super continuum generation using ultra short pulse fiber laser. 978-1-4244-2611 IEEE
11. Herrmann J, Griebner U, Zhavoronkov N, Husakou A, Nickel D, Knight JC, Wadsworth WJ, Russell PStJ, Korn G (2002) Experimental evidence for supercontinuum generation by fission of higher-order solitons in photonic fibers. Nature 424:847–851
12. Gonzalez-Herraez M, Martin-Lopez S, Corredera P, Hernanz ML, Horche PR (2003) Supercontinuum generation using a continuous-wave Raman fiber laser. Opt Commun 226:323–328
13. Foster MA, Gaeta AL, Dudley JM, Cao Q, Lee D, Trebino R (2005) Supercontinuum generation and pulse compression in sub-wavelength-sized waveguides. Conference on Lasers and Electro-Optics (CLEO), pp 1261–1263
14. Chow KK, Takushima Y, Mizuno Y (2006) High average power super-continuum generation using a 1-μm ASE noise source. Optical Society of America
15. Dou L, Gao Y, Xu A, Tang M, Shum P (2002) Super-continuum generation using noise-like pulses from a large normal dispersion passively mode locking fiber laser. IEEE 92–93
16. Dekker S, Xiong C, Magi E, Judge AC, Sanghera JS, Shaw LB, Aggarwal ID, Moss DJ, Eggleton BJ (2010) Broadband low power super-continuum generation in As2S3 chalcogenide glass fiber nanotapers. Optical Society of America

17. Snyder AW, Love JD (2000) Optical waveguide theory. Kluwer Academic, Dordrecht
18. Birks TA, Knight JC, St P, Russell J (1997) Endlessly single-mode photonic crystal fiber. Opt Lett 22:961–963
19. Poli F, Cucinotta A, Selleri S (2007) Photonic crystal fibers. Springer, Berlin
20. Agrawal GP (2000) Non-linear fiber optics. 4th edn. Springer, Berlin, pp 471–480
21. Dudley JM, Genty G, Eggleton BJ (2008) Harnessing and control of optical rogue waves in supercontinuum generation
22. Chen HW, Chen SP, Hou J (2011) 7 W all-fiber supercontinuum source. Laser Phys 21(1):191–193 ISSN 1054_660X
23. Walewski JW, Filipa JA, Hagen CL, Sanders ST (2006) Standard single-mode fibers as convenient means for the generation of ultrafast high-pulse-energy super-continua. Appl Phys B 83:75–79
24. Nishizawa N, Hori M (2007) Super continuum generation using ps high energy erdoped fiber laser at 1.55 μm. The 7th pacific rim conference on lasers and electro-optics (CLEO/Pacific Rim 2007)
25. Travers JC, Rulkov AB, Cumberland BA, Popov SV, Taylor JR (2008) Visible supercontinuum generation in photonic crystal fibers with a 400 W continuous wave fiber laser. Optical Society of America
26. Kobtsev SM, Smirnov SV (2005) Modeling of high-power supercontinuum generation in highly nonlinear, dispersion shifted fibers at CW pump. Opt Express 13(18):6912–6918 5 Sept 2005
27. Kachalovaa NM, Voitsekhovich VS, Borodina AM, Khomenko VV, Pentegov SY (2011) Femtosecond supercontinuum characteristics control. Opt Spectrosc 111(4):593–598 ISSN 0030400X
28. Kibler B, Fischer R, Genty G, Neshev DN, Dudley JM (2008) Simultaneous fs pulse spectral broadening and third harmonic generation in highly nonlinear fiber: experiments and simulations. Appl Phys B 91:349–352
29. Kobtsev SM, Kukarin SV (2009) Spectral broadening of femtosecond pulses in an nonlinear optical fiber amplifier. Opt Spectrosc 107(3):344–346 ISSN 0030_400X

Chapter 7
All-Optical Frequency Shifting

Abstract This chapter provides a theoretical model for pulse propagation inside an SOI waveguide. The four-wave mixing (FWM) process in SOI waveguides is discussed with an emphasis on the effects of two-photon absorption and the consequent free-carrier effects. All-optical wavelength conversion and optical signal processing along with requisite devices are illustrated with examples exclusively.

Keywords Frequency shifting • All-optical wavelength conversion

Abbreviations

SOI	Silicon on insulator
FWM	Four-wave mixing
WDM	Wavelength division multiplexing
DWDM	Dense wavelength division multiplexing
OTDM	Optical time division multiplexing
MZ	Mech-Zehnder
GVD	Group velocity dispersion
OXC	Optical cross-connect
CW	Continuous wave
MEMS	Micro-electromechanical system
TPA	Two-photon absorption
FCA	Free carrier absorption
SPM	Self-phase modulation

J. Ahmed et al., *Optical Signal Processing by Silicon Photonics*, SpringerBriefs in Materials, 81
DOI: 10.1007/978-981-4560-11-5_7, © The Author(s) 2013

7.1 Introduction

Several active and passive SOI waveguide-based devices have recently emerged
in the market. The important active devices are optical amplifiers, modulators,
optical lasers, switches, optical detectors, logic gates, etc. On the other hand, passive devices including wavelength division multiplexers, splitters, interference
couplers, Mach-zenhder-interferometer-based filters, optical delay lines, switch
matrix, etc., are the necessary building blocks of a complete monolithic integration
system on the silicon platform.

7.2 All-Optical Wavelength Conversion

All-optical wavelength conversion is the most frequently required phenomenon
in present and future wavelength division multiplexing (WDM) transmission systems. The utilization of multiplexing techniques such as dense wavelength division
multiplexing (DWDM) and optical time division multiplexing (OTDM) has shown
an enormous increase in the optical transmission capacity. WDM systems with a
capacity of 1.6 Tb/s (by multiplexing 160 channels at 10 Gb/s) are commercially
available and transmission systems having capacity of >6.4 Tb/s (i.e. 160 channels at 40 Gb/s) are on the way. Now, it is the need of the moment to achieve high
modulation bit rates, effective signal routing, and switching in the optical domain.
In silicon waveguides these wavelength conversions can be achieved by exploiting the nonlinearities of the waveguides [1, 2], free carrier generation-based crossabsorption modulation, and four-wave mixing (FWM) [3]. The focus is mainly the
wavelength conversion in silicon waveguides based on FWM (Fig. 7.1).

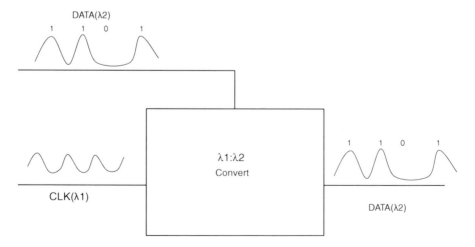

Fig. 7.1 Simple wavelength conversion scheme [2]

The California Institute of Technology presented the frequency conversion with nonlinear optical polymers and high index contrast waveguides, and patented their invention relating to slotted silicon/silicon dioxide waveguide for manipulating light. The apparatus comprises a substrate (silicon wafer) having an insulating surface (silicon and oxygen), a high index contrast (silicon) waveguide adjacent to the insulating surface, and a cladding adjacent to high index contrast waveguide comprising a material that exhibits an enhanced nonlinear coefficient. The high index contrast waveguide and cladding are configured so that an electric field intensity of at least 105 V/m is generated in response to an optical input of not more than 1 mW of continuous wave (CW) light [4].

Parametric coupling enabled by Raman susceptibility is used by V. Raghunathan et al. to demonstrate wavelength conversion in silicon waveguides. Silicon has the gain coefficient for Stimulated Raman Scattering that is -10^4 times higher than that in glass fibers, and the high index contrast between silicon and SiO_2/air provides the possibility to meet signal amplification and wavelength conversion in the chip-scale. Raman susceptibility can be used to transfer optical information between the Stokes and anti-Stokes frequency channels; this conversion efficiency is dependent upon:

1. the intensity of the pump wave,
2. the Raman susceptibility of silicon, and
3. the phase difference between the waves [5].

The wavelength conversion techniques used in $LiNbO_3$ crystal-based wavelength converters have drawbacks such as photorefractive damage and high cost. To overcome these sufferings, Haisheng Rong and Mario J.Paniccia from Intel Corporation came up with their design which uses semiconductor optical waveguide. They employed the concept of FWM process to realize wavelength conversion in semiconductor optical waveguides with a p-i-n diode structure. FWM is an inter-modulation phenomenon in nonlinear optical systems. When three wavelengths (λ_1, λ_2, and λ_3) interact in a nonlinear medium, they give rise to a fourth wavelength (λ_4) which is established by the scattering of the incident photons, generating the fourth photon. FWM is also in attendance if only three wavelengths interact, i.e., $\lambda_0 = \lambda_1 + \lambda_1 - \lambda_2$ which is referred to as degenerate FWM, showing similar properties as four interacting waves. A general spectrum of FWM signals for high speed all-optical wavelength converters illustrate a relationship between the pump (λ_1), input (λ_2), and converted (λ_3) signals [6].

Mark Foster et al. published their invention which is a silicon integrated photonic optical parametric amplifier, oscillator, and wavelength converter. Their device generally relates to optical signal processing and amplification, and more specifically to the optical transmission systems for wavelength shifting, phase matching, and amplification of optical signals performed by a semiconductor photonic device.

In existing telecom transmission systems, where optical information is transmitted via optical fibers, degradation of optical signal, such as pulse dispersion is created by the medium (material dispersion), which causes spreading of an optical signal pulse traveling through the fiber. When this optical signal is dispersed in frequency, some signal wavelengths travel faster than others of the same signal. Another type of dispersion is dispersion which further degrades the signal due to signal spreading induced by the geometries and waveguide the dimensions of optical waveguides. These are the limitations of the efficiency of transmission systems.

The invention by Mark Foster et al. presents a solution for optical signal intensity adjustment or light signal amplification, optical signal wavelength, and phase adjustment. The silicon waveguide used in this invention provides anomalous group velocity dispersion within the range of a few picoseconds per nanometer of wavelength and kilometer of distance (about 1,856 picoseconds per nanometer of wavelength and kilometer of distance), which is well suited for frequency mixing and in turn for wavelength conversion [7].

By using high optical nonlinearity and light confinement offered by silicon waveguides, Haisheng Rong et al. from Intel Corporation have demonstrated their chip- scale Raman laser, amplifier, and wavelength converter. An 8-cm long silicon p-i-n waveguide was selected for FWM occurrence and wavelength conversion was successfully demonstrated with input signal carrying information at 40 Gb/s PRBS data [8] (Fig. 7.2).

A broadband optical gain and wavelength conversion can be made more efficient by precisely selecting the waveguide dimensions and proper phase matching. The amplification by FWM is critically dependent on the phase mismatch among the pump, signal, and idler waves [9].

Mark A Foster et al. demonstrated the broad-bandwidth optical gain and wavelength conversion in silicon waveguides using phase-matched FWM in SOI channel waveguides pumping with a single laser at 1550 nm, achieving wavelength conversion from 1,511 to 1,591 nm, over a 29 nm range with peak gain of 13.9 dB for net gains as high as 4.2 dB. They used the suitable design of the waveguides to produce anomalous group-velocity dispersion (GVD) in this wavelength regime [10].

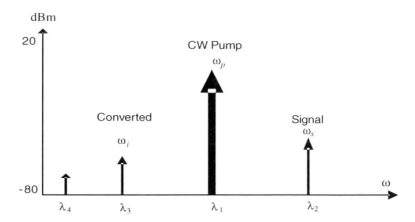

Fig. 7.2 Spectrum of the FWM signals

Wavelength conversion over the entire ITU S, C, and L bands at a data rate of 40 Gb/s is demonstrated using silicon photonic waveguides, a continuous-wave pump, and a non-return-to-zero (NRZ) input signal of wavelength 1,513.7 nm is up-converted across nearly 50 nm, resulting in a minimum power penalty of 2.9 dB on the converted signal at a BER of 10^{-9} [11].

160-Gb/s pulsed return-to-zero (RZ) wavelength conversion within a CMOS-compatible silicon chip device comprising 1.1 cm long and a 290-nm × 660-nm cross section has been published recently which is the highest data rate achieved till date for a single-channel conversion in silicon, with a 21-nm conversion range and −15.5 dB conversion efficiency using continuous-wave (CW) pump producing moderate pulse broadening, Fig. 5.2 [12].

The new schemes were developed during the 1990s for making wavelength converters; in the present technology the wavelength converters change the input wavelength to a new wavelength without modifying the data content of the signal.

A simple scheme uses an optoelectronic regenerator; this scheme is relatively easy to implement as it uses standard components. In this scheme an optical receiver first converts the incident signal at the input wavelength λ_1 into an electrical bit pattern, which is then used by a simple transmitter to generate the optical signal at the desired wavelength λ_2. Advantages include insensitivity to the input polarization while the drawback includes limited transparency to bit rate and data format and speed limited by electronics [13].

7.3 Optical Modulation

Nowadays, silicon-photonic chip-scale active and passive devices are a hot and overwhelming topic for researchers in this field. The devices being proposed and demonstrated as a result of this titanic research and evolution of nanofabrication techniques incorporate silicon lasers, amplifiers, modulators, photo-detectors, wavelength converters, optical logic gates, optical buffers, biosensors, etc. The list of silicon photonic devices is increasing and applications of these nanoscale devices will continue at a great pace [14–16]. The efficient fiber to silicon-waveguide coupling (from cross-sectional dimensions of several micrometers to a few hundreds of nanometers in centimeter lengths) enables the better use of silicon waveguides.

Waveguides and devices with enhanced third-order nonlinearities in polymer silicon which is essential for nanometer scale photonic signal processing devices have been patented by researchers at the University of Washington [17]. They used these slot waveguides having closed as well as linear or circuitous formations in devices, i.e., variable delay lines, optical logic gates, optical multiplexers, optical self-oscillators, and optical clock generators. Passive silicon devices such as splitters, bends, couplers, and filters have been designed but the signal cannot be modulated while propagating through them [18, 19]. Optical switches and modulators based on silicon and using concentrations of free-carriers have been demonstrated which require high powered pump light beams to gain modulation [20–23]. The reason behind this phenomenon is the weak complex-refractive-index drift from the mean position as a function of concentration of free carriers [24]. However, the use of resonators having strong optical light signal confinement can alleviate the manufacturing of these devices.

The optical modulators at bit rates of 10 Gb/s or higher, the frequency chip imposed by direct modulation becomes large enough that direct modulation of semiconductor lasers is rarely used. For such high speed transmitters the laser is biased at a constant current to provide the CW output [25]. The optical modulator is placed next to the laser converter, the CW light into a data-coded pulse train with the right modulation format.

The two types of optical modulators developed for the light wave system applications are:

1. $LiNbO_3$ modulator in the Mach- Zehnder configuration and
2. Semiconductor modulator-based on electro absorption.

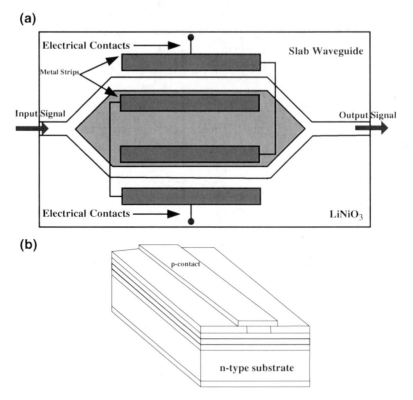

Fig. 7.3 a $LiNbO_3$ modulator in the Mach-Zehnder configuration (external modulator), **b** Semiconductor modulator based on electro-absorption [27]

An optical modulator makes use of the LiNbO$_3$ material and a Mech-Zehnder (MZ) interferometer for intensity modulation. The refractive index of electro-optic materials such as LiNbO$_3$ can be changed by applying external voltage. LiNbO$_3$ modulators with a bandwidth of 10 GHz were available commercially by 1998, and the bandwidth increased to 40 GHz by 2000. The second category of optical modulator is electro absorption modulators, that they made by using the same semiconductor material that is used for the laser; thus the two can be easily integrated on the same chip. The electro absorption modulator makes use of the Franz-Keldysh effect in which the band gap of a semiconductor decreases when an electric field is applied across it [26].

Other materials can also be used to make the external modulators, for example, the modulators have been fabricated using electro-optic polymers; such modulators may find applications in light wave systems (Figs. 7.3).

Hewlett Packard Company also patented their design of optical modulator in 2009 including electrically controlled ring resonator, comprising:

1. a waveguide for guiding an optical signal,
2. a ring resonator disposed in evanescent communication with the waveguide for at least one predetermined wavelength of the optical signal, and
3. a semiconductor PN junction which receives an electrical control signal.

A resonance state of the ring resonator at the predetermined wavelength is thus controlled by the electrical signal, which controls a free carrier population in the resonant light path, and the optical signal is thereby modulated according to the electrical control signal [28]. A similar high speed ring optical modulator is patented by the Electronics and Telecommunications Research Institute, Daejeon [29].

7.4 All-Optical Switching

All-optical switching and interconnection based on SOI are fitting the need of the hour due to their promise of fulfilling the ultrahigh bit rates of hundreds of Tbits/s inside and between computer chips. These computer chips will contain about 1,000 cores and more than 10 tera flip-flops in 2020 [30]. Many schemes have been developed for performing the switching operations, mechanical switching is the simplest to understand. A micro-electro-mechanical system (MEMS) is also used for switching. Semiconductor waveguides can also be used for making the optical switches in the form of direction couplers, MZ interferometers, or Y junctions [31].

The optical cross-connect (OXC) performs the same function as provided by electronic digital switches in telephone networks. The design and fabrication of OXCs has remained a major topic of research since the arrival of WDM systems [32].

The future computer architecture is looking curiously toward silicon photonics for mature solution but the on-chip integration of all-optical silicon-based switching is challenging due to its weak nonlinear optical properties. The switching in silicon can be achieved at the cost of very high power intensities but again are not suitable for nanoscale chips.

Almeida et al. [33] presented the experimental demonstration of efficient optical switching in silicon resonant structures (Fig. 4.3) by strong light-confining to enhance the sensitivity of light to small changes in refractive index. They claimed that the transmission efficiency of the structure can be modulated by up to 94 % in less than 500-ps of switching time using light pulses with energies as low as 25 pJ. They used ring resonator of 10 μm diameter, while both the silicon waveguide and the ring resonator are channel waveguides with 450-nm-wide by 250-nm-high rectangular cross-sections.

Michael Forst et al. demonstrated high-speed all-optical switching via vertical excitation of electron–hole plasma in oxygen-ion implanted silicon on insulator micro-ring resonator. The spectral response of the device is rapidly modulated by photo-injection and subsequent recombination of charge carriers at artificially introduced fast recombination centers, at an implantation dose of 1×10^{12} cm^{-2}. The carrier's lifetime is reduced to 55 ps, which facilitates optical switching in the 1.55 μm wavelength range at modulation speeds of more than 5 Gbits/s [34].

Jeffery J. Maki reported optical switches; one consisting of two waveguides crossing each other at an angle to form an intersection, and a pair of electrodes placed within a proximity of the intersection to switch a light traveling from the first waveguide to the second waveguide. This intersection includes a geometry that supports single and multimode propagation (the intersection includes a geometry having a ridge width ranging approximately from 2.6 to 19.2 μm and a ridge height ranging approximately from 4 to 16 μm) [35].

In [36] a waveguide-type optical switch based on amplifier is disclosed, which amplifies an optical signal passing through this waveguide (by performing an optical pumping through a WDM optical coupler in the said optical waveguide, comprising a fluorescence emitting electro-optic material), and at the same time carries out an optical switching by using an optical waveguide refraction index change induced by an electro-optic effect under an electrical control (Fig. 7.4).

Fig. 7.4 Silicon resonant structure [32]

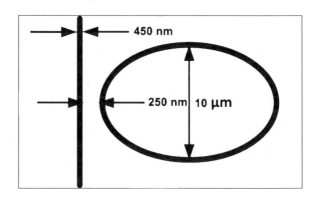

7.5 Optical Sources (Laser) and Amplifiers

Due to indirect band gap, silicon has very weak optical emission cross-section and thus is not considered as an appropriate material to be used in optical sources; so is the absent brick in the monolithic integration. Researchers in different R&D organizations and in education-based research projects are working on silicon light emitters that have been most challenging so far. The light generation in semiconductors is a consequence of electron–hole pair recombination. The silicon recombination process needs a phonon for momentum conservation due to its indirect energy band gap which minimizes the chance of luminescence [37] (Fig. 7.5).

Different approaches are being used to overcome the silicon constraints in emitting light, i.e., controlling the free carrier lifetime by quantum confinement, Raman amplification for emission, or doping with other rare earth materials. Silicon-based lasers can contain layers of SiGe/Si in the active region while silicon is an indirect band gap material. Germanium is nearly a direct band gap material, so a quantum well structure with a Ge well and a SiGe barrier is a strong candidate for realizing a room-temperature electrically pumped group IV laser diode operating at, or near, the fiber-optic communications wavelengths [30].

UCLA [38] and Intel [39] have presented the first silicon on-chip light emitters (Fig. 5.5) using a CW pump. Previously, lasing was done by pulsed pumping; also, in this invention a reverse biased p-i-n diode was used to reduce the two-photonic absorption (TPA) and induced free carrier absorption (FCA) in silicon (Fig. 7.6).

Overall amplification or net gain of an optical signal is reachable in silicon now, but careful active carrier removal in Raman amplifiers and lasers is needed. In Nicolaescu and Paniccia [40] Intel disclosed a Raman ring resonator-based laser and wavelength converter which includes a pump directing optical beam (λ_p) into a ring resonator; emission of an optical beam (λ_s) is caused in the resonator by propagating the pump optical beam around the ring resonator. The pump power level is to be sufficient to cause the emission of the second optical beam. The pump optical beam is directed out of the resonator after a round trip, the second optical beam is re-circulated around the resonator to further stimulate the emission. A better semiconductor-based Raman laser and/or amplifier with reduced TPA generated carrier lifetime was patented by the same in 2007, the diode

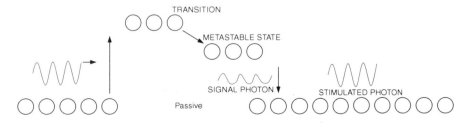

Fig. 7.5 Light signal amplification process [30]

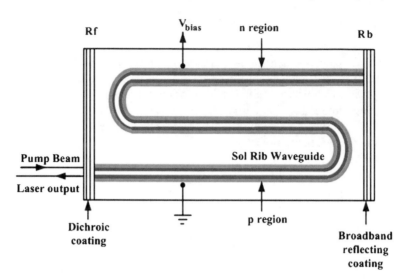

Fig. 7.6 Silicon resonant structure [39]

structure is to be incorporated and biased to sweep out free carriers from the optical waveguide generated in response to TPA in the optical waveguide [41].

Stimulated Raman scattering (SRS) effect at infrared frequencies was discovered in 1962 and TPA process with a full quantum mechanical calculation was described afterwards to account for anti-Stokes generation and higher order Raman effects, while coupled-wave formalism was adopted to describe the stimulated Raman-effect. Self-focusing was later included to give a reason for the much larger gain observed in SRS which facilitated the study and design of Raman amplifiers and lasers. The Trustees of Columbia University patented their design in [42, 43].

There is an increasing demand for tunable lasers fueled by the advent of WDM which has become widespread in fiber optic communication systems. In [44] Light wire formed an SOI based tunable laser, consisting of a tunable wavelength reflecting element and associated phase matching elements, these are formed on the surface of the SOI structure, with optical waveguides shaped in the SOI surface layer providing communication between these components. The tunable wavelength element is controlled to adjust the optical wavelength. Macquarie University, New South Wales patented the design of an optical amplifier laser [45], on a similar principle.

7.6 Optical Parametric Amplifier

An optical parametric amplifier, abbreviated as OPA, is a laser light source that emits light of variable wavelengths by an optical parametric amplification (OPA) process.

7.6.1 Optical Parametric Generation

This light emission is based on the nonlinear optical principle. By a nonlinear optical crystal, the photon of an incident laser pulse (pump) is divided into two photons. The sum energy of these two photons is equivalent to the energy of the photon of the pump. Ordinary and extraordinary polarizations generate light; the ordinary light is called the signal and the extraordinary light is called the idler. By phase matching condition, the wavelengths of the signal and the idler are determined. The wavelength is changed by the angle between the incident pump laser ray and the axes of the crystal. By changing the phase matching condition, the wavelengths of the signal and the idler lights can be tuned. This process is termed as optical parametric generation or OPG.

7.6.2 Optical Parametric Amplification

After separation of the signal light from the OPG outputs, the remaining idler passes through a nonlinear optical crystal collinearly with light of the same wavelength as the pump, and a stronger output of the same wavelength as the signal and idler is acquired as the output of the OPA. These wavelength-variable outputs are efficiently used in many spectroscopic methods. As an example of OPA, the incident pump pulse is the 800 nm (12,500 cm^{-1}) output of a Ti:sapphire laser, and the two outputs, signal and idler, are in the near-infrared region, the sum of the wave number of which is equal to 12,500 cm^{-1}.

7.7 Raman Amplification

Usually, Raman amplification is based on the SRS phenomenon. In the nonlinear regime when a lower frequency 'signal' photon that induces the inelastic scattering of a higher frequency 'pump' photon in an optical medium, as a result, another 'signal' photon is produced with the surplus energy resonantly that passed to the vibrational states of the medium. This procedure allows all-optical amplification as with other stimulated emission processes. Nowadays, mostly optical fiber is used as the nonlinear medium for SRS for telecommunication systems. In this case, it is characterized by a resonance frequency downshift of ~11 THz (corresponding to a wavelength shift at ~1,550 nm of ~90 nm). The SRS amplification process can be readily cascaded, thus accessing essentially any wavelength in the fiber low-loss guiding windows (both 1,300 and 1,550). Raman amplification is used in optical telecommunications allowing all-band wavelength coverage and in-line distributed signal amplification in addition to applications in nonlinear and ultrafast optics.

7.8 Stokes and Anti-Stokes Scattering

Basically, there are two types of Raman scattering, i.e., Stokes scattering and anti-Stokes scattering. The different possibilities of visual light scattering are as follow:

- Rayleigh scattering (no Raman effect),
- Stokes scattering in which molecule absorbs energy, and
- Anti-Stokes scattering in which molecule loses energy.

7.8.1 Stimulated Scattering and Amplification

Raman amplification can be obtained by using SRS, which actually is a combination of a Raman process with stimulated emission. It is interesting for application in telecommunication fibers to amplify inside the standard material with low noise for the amplification process. However, the process requires significant power and thus imposes more stringent limits on the material. The amplification band can be up to 100 nm broad, depending on the availability of allowed photon states.

7.8.2 Spectrum Generation

For high intensity CW lasers, SRS can be used to produce broad bandwidth spectra. This process can also be seen as a special case of FWM, where the frequencies of the two incident photons are equal and the emitted spectra are found in two bands separated from the incident light by the phonon energies. The initial Raman spectrum is built-up with spontaneous emission and is amplified later on. At high pumping levels in long fibers, higher order Raman spectra can be generated by using the Raman spectrum as a new starting point, thereby building a chain of new spectra with decreasing amplitude. The disadvantage of intrinsic noise due to initial spontaneous process can be overcome by seeding a spectrum at the beginning, or even using a feedback loop like in a resonator to stabilize the process. Since this technology easily fits into the fast evolving fiber laser field and there is demand for transversal coherent high intensity light sources (i.e., broadband telecommunication, imaging applications), Raman amplification and spectrum generation might be widely used in the near future.

7.9 Brillouin Scattering

Brillouin scattering, named after Léon Brillouin, which occurs when light in a medium such as water or a crystal interacts with time-dependent density variations and changes its energy (frequency) and path. The density variations may be due to acoustic modes such as phonons, magnetic modes like magnons, or temperature gradients. As described in classical physics, when the medium is compressed its index of refraction changes and the light's path necessarily bends.

7.9.1 Relationship to Raman Scattering

Brillouin scattering is related to Raman scattering in a way that both represent light's inelastic scattering processes with quasi-particles. The distinction exists in the type of information extracted from the sample and the detected range of frequency shift. Brillouin scattering denominates the photons scattering from quasi-particles while in case of Raman scattering, interaction with vibrational and rotational transitions in molecules makes photon scatter. Therefore, the information provided by these two techniques about the sample is different. Raman spectroscopy is used to determine the molecular structure and chemical composition, while Brillouin scattering is capable of measuring the properties on a larger scale such as the elastic behavior. From an experimental point of view, in Brillouin scattering the frequency shifts are detected with an interferometer while Raman setup can be based on either dispersive (grating) spectrometer or interferometer.

7.9.2 Stimulated Brillouin Scattering

For intense beams (e.g. laser light) traveling in a medium such as an optical fiber, acoustic vibrations in the medium is created due to variations in the electric field of the beam itself through electrostriction. As a result of these vibrations, the beam might endure Brillouin scattering, which usually occurs in the direction opposite to the incoming beam, a phenomenon known as stimulated Brillouin scattering. For gases and liquids, the frequency shifts typically are of the order of 1–10 GHz. The effect by which optical phase conjugation can take place is Stimulated Brillouin scattering.

7.10 XPM or Cross-Phase Modulation

Generally, the cross-phase modulation or XPM is an optical upshot nonlinear in nature, in which the phase of wavelengths can get effected through the optical Kerr effect. This implies the modulation technique that can be used for the addition of a light stream to the information by modification of the coherent optical beam phase with another beam through interactions in a proper nonlinear medium, which is cross-phase modulation. This method is functional to optical fiber communications for various applications. In DWDM applications with direct detection is an intensity modulation (IM-DD).The XPM effect is a process having two steps: First, the signal is modulated when its phase modulation is done by the co-propagating second signal, whereas in the second step, transformation of the phase modulation into a power variation dispersion leads to a. Further, the cut-off connection can occur between the channels due to dispersion and thereby XPM-effect is reduced.

7.11 Self-Phase Modulation

Self-phase modulation, aka SPM, is defined as a nonlinear optical effect of interaction of light and matter. A varying refractive index of the medium due to the optical Kerr effect will induce when an ultrashort pulse of light is traveling in a medium. This refractive index variation will produce a phase shift in the pulse. This phase shift also leads to a change in the pulse's frequency spectrum. Self-phase modulation is an important effect in optical systems that uses short intense pulses of light, for example, laser and optical fiber communications systems.

7.12 Free Carrier Absorption

Theoretically, FCA occurs when a material absorbs a photon and a carrier is excited from a filled state to an unoccupied state in the same band. FCA is very different from interband absorption in semiconductors. The reason behind this is that the electron being excited is a conduction electron that can move freely, hence FCA.

7.13 Optical Detectors

A monolithically scaled optical interconnect system based on SOI technology requires four major components:

- A light source,
- An optical modulator for encoding data into light pulses,
- A waveguide for efficiently transporting light across the chip, and
- A photodetector for converting optical signals into electrical signals.

One of the necessary components of an optical fiber communication system is decoder, which is also a fundamental element dictating the overall performance of the system. The function it performs is converting the received signal from optical to electrical, then this signal is amplified before processing it further. A detector must satisfy the criteria and stability of performance characteristics.

   The lack of an effective photodetector for such applications has affected the development of scalable SOI interconnections. For CMOS-compatible processing the photodetector must be made from silicon (Si) or germanium (Ge), but Si is transparent from wavelength range of 1.1–1.8 μ, hence if Si waveguide is to be used then we have to focus on this range only. Using Ge is advantageous but the refractive index gap between Si and Ge makes it unsuitable for a monolithically integrated Si photonics. Ge is also suffered by its large dark current, impaired bandwidth, multimode nature, and inefficient waveguide to detector coupling. Germanium or GeSi on Si provides detection of optical signals at 1.3–1.5 μm. Luxtera demonstrated the Germanium-enabled SOI CMOS process and Intel produced an avalanche photodiode of Ge and Si [46, 47].

   A silicon-waveguide photodetector is exposed, comprising a waveguide layer where only single mode light of a fixed polarization is guided/confined, and a detection layer formed on the waveguide layer where guided light is detected [48]. Luxtera patented a germanium on silicon waveguide photodetector (disposed on SOI substrate) which generates an electric current when an infrared optical signal travels through the photodetector [49].

   Michael et al. designed a silicon photo-detector device comprising a nanoscale silicon high-contrast waveguide, an optical input, and an electrical output. The distribution across the waveguide in optical mode has peak intensity in correspondence to the nanoscale silicon waveguide's surface states [50]. In Fig. 5.6 an equivalent designed structure's electrical circuit is shown (Fig. 7.7).

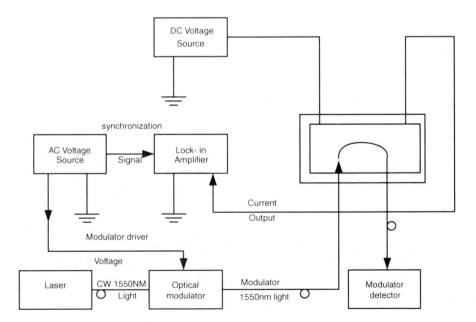

Fig. 7.7   An equivalent electrical circuit of the designed structure [49]

7.14 Other Optical Signal Processing Chip-Scale Devices

As the SOI has offered a promising platform for complete monolithic integration of Si-photonic devices, numerous signal processing devices have been constructed recently. Optical filters are designed in [51, 52] but most suffer from polarization dependence due to structural birefringence offered by the waveguides. Transmission spectrum of the ring resonator filter with polarization diversity has designed which works well for both polarizations (TE and TM), and polarization-dependent loss is about 1 dB [53].Optical splitters, attenuators, pulse shapers, optical delay lines, optical logic gates (AND, NOT, NAND, NOR, OR, XOR, and XNOR), etc., have been successfully demonstrated using nonlinear properties of silicon.

Chip-scale optical delay lines (buffers) were proposed using slow light propagation [54], or coupled resonator waveguide [55] and cascaded microring resonator by IBM Inc., which consists of 100 resonators producing about 200 ps delay. A novel-shaped microdisk resonator is presented to overcome evanescent coupling through a narrow gap and offers gap-less nonevanescent light coupling with mode matching [56]. An SOI-based variable optical attenuator (VOA) is designed in which propagating light is absorbed by the injected carriers and shows nanosecond temporal response [57].

In Poon and Xu [58] a 5 × 5 Si photonic optical router is demonstrated comprising electro-optic switchable micro-ring resonators in cascaded form, coupled tangentially into a network of cross-grid waveform, to lessen the insertion loss due to waveguide crossing and decrease the leakage to the crossed waveguides; waveguide crossings based on MMI is used. An SOI-based optical interconnection arrangement is provided that significantly reduces the size, complexity, and power consumption of conventional high density electrical interconnections. A group of optical modulators and wavelength division multiplexers/demultiplexers are used

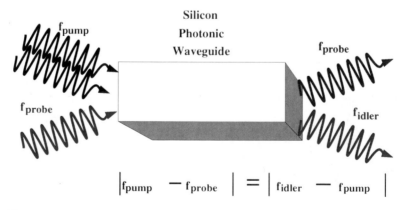

Fig. 7.8 FWM, using two pump waves in 3R regeneration [61]

in alliance with traditional electrical signal paths to "concentrate" a large number of the electrical-pinouts onto one optical waveguide [59].

To cope with the light signal deterioration due to miscellaneous optical components during propagation, optical reshaping, reamplification (2R), and retiming (3R) is necessary. Generally, a CW pump is used in 2R and a clocked modulated (CM) pump is employed for 3R. Higher order four-wave mixing (HO-FWM) was used for 3R regeneration in [60] with a CM pump. Ozdal Boyraz demonstrated nanoscale signal regeneration using nonlinear phenomenon of FWM that has already been discussed. Gaete et al. demonstrated different techniques in FWM to improve 2R and extinction ratio. In one experiment an input pump signal at 1,552 nm with 1,546 nm probe combines to generate 1,558 nm idler, extinction ratio increased was 4.2 dB at about 100 mW pump signal intensity.

The generated idler has the timing information about the input signal but no timing jitter improved. With the use of CM pump [61] instead of CW the extinction ratio can be improved along with timing jitter reduction in the matter of nanoscale structures (Fig. 7.8).

References

1. Claps R, Raghunathan V, Dimitropoulos D, Jalali B (2003) Anti-stokes Raman conversion in silicon waveguides. Opt Express 11:2862–2872
2. Raghunathan V, Claps R, Dimitropoulos D, Jalali B (2004) Wavelength conversion in silicon using Raman induced four-wave mixing. App Phys Letts 85:34–36
3. Fukuda H, Yamada K, Shoji T et al (2005) Four-wave mixing in silicon, wire waveguides. Opt Express 13:4629–4637
4. Hochberg M, Baehr-Jones T (2008) US20080002992
5. Raghunathan V, Dimitropoulos D, Claps R, Jalali B (2003) Wavelength conversion in silicon waveguides using parametric Raman coupling. Opt Express 11:2862–2872
6. Rong H, Paniccia MJ (2007) US007256929
7. Foster MA, Gaeta AL, Lipson M, Sharping JE, Turner A (2009) US20090060527
8. Rong H, Xu S, Ayotte S, Cohen O, Raday O, Paniccia M (2008) Silicon based chip-scale nonlinear optical devices: laser, amplifier, and wavelength converter. LEOS winter topical meeting, Sorrento, Italy
9. Hansryd J, Andrekson A, Westlund M, Li J, Hedekvist P (2002) Fiber-based optical parametric amplifiers and their applications. Quantum Electron 8:506–520
10. Foster MA, Sharping JE, Gaeta AL, Turner AC, Schmidt BS, Lipson M (2006) Broadbandwidth optical gain and efficient wavelength conversion in silicon waveguides. Conference on laser and electro-optics, Long Beach, Canada
11. Lee BG, Biberman A, Amy C et al (2009) Demonstration of broadband wavelength conversion at 40 Gb/s in silicon waveguides. Photonics Technol Lett 21:182–184
12. Lee BG, Biberman A, Amy C et al (2009) 160-Gb/s broadband wavelength conversion on chip using dispersion-engineered silicon waveguides. Conference on laser and electro-optics, Baltimore, USA
13. Duan GH (1995) In: Agrawal GP (ed) Semiconductor lasers: past, present, and future. AIP Press, Woodbury
14. Claps R, Raghunathan V, Dimitropoulos D, Jalali B (2004) Influence of nonlinear absorption on Raman amplification in silicon waveguides. Opt Express 12:2774–2780

15. Xia F, Sekaric L, Vlasov Y (2007) Ultra compact optical buffers on a silicon chip. Nat Photonics 1:65–71
16. Vos KD, Bartolozzi I, Schacht E, Bienstman P, Baets R (2007) Silicon-on-insulator microring resonator for sensitive and label-free bio-sensing. Opt Express 15:7610–7615
17. Jones BT, Michael HJ (2009) WO2009111610
18. Loncar M, Doll T, Vuckovic J, Scherer A (2000) Design and fabrication of silicon photonic crystal optical waveguides. Lightwave Technol 18:1402–1411
19. Lin MS, Chou CM (2009) US7582966
20. Almeida VR, Barrios CA, Panepucci RR (2004) All-optical switching on a silicon chip. Opt Lett 29:2867–2869
21. Dinu M, Quochi F, Garcia H (2003) Third-order nonlinearities in silicon at telecom wavelengths. App Phys Letts 82:2954–2956
22. Soref RA, Bennett BR (1987) Electro-optical effects in silicon. Quantum Electron 23:123–129
23. Alexandre B, Kamins TI (2009) US20090190875
24. Soref RA, Bennett BR (1987) Electro-optical effects in silicon. Quantum Electron 23:123–129
25. Arawal GP (2008) Fiber-optic communication systems, 3rd edn. Academic Press, Boston
26. Howerton MM, Moeller RP, Greenlatt AS, Krahenbuhl R (2000) Fully packaged, broad-band LiNbO3 modulator with low drive voltage. IEEE Photon Technol Lett 12:792–794
27. Harry Dutton (1998) Understanding optical communications. IBM Corporation
28. Alexandre B, Kamins TI (2009) US20090190875
29. Park JW, Kim, Kim G, Kim HS, Mheen B (2010) US7646942
30. Yoo SJB (2009) Future prospects of silicon photonics in next generation communication and computing systems. Elect letters 45:584–588
31. Reanud M, Bachmann M, Ermann M (1996) Semiconductor optical space switches. IEEE J Sel Topics Quantum Electron 2:277–288
32. Mutafungwa E (2001) An improved all-fiber cross-connect node for future optical transport networks. Opt Fiber Technol 7:236–253
33. Qianfan Xu, Michal Lipson (2007) All-optical logic based on silicon micro-ring resonators. Optics Express 15(3):924–929
34. Först M, Niehusmann J, Plötzing T et al (2007) High-speed all-optical switching in ion-implanted silicon-on-insulator microring resonators. Opt Letts 32:2046–2048
35. Maki JJ (2009) US20090310910
36. Kim KH, Choi YG, Lee HK (2003) US6538804
37. Hammond RB, Silver RN (1980) Temperature dependence of the exciton lifetime in high-purity silicon. App Phys Letts 36:68–71
38. Jalali B, Raghunathan V, Dimitropoulos D, Boyraz O (2006) Raman based silicon photonics. Quantum Electron 12:412–421
39. Rong HS, Jones R, Liu AS et al (2005) A continuous-wave Raman silicon laser. Nature 433:725–728
40. Nicolaescu R, Paniccia MJ (2006) US7046714
41. Liu A, Paniccia MJ, Rong H (2007) US7266258
42. Wei WC, Xiaodong Y (2006) WO2006014346
43. Wei WC, Xiaodong Y (2009) US7532656
44. Mark W, Piede D, Prakash G (2009) US20090135861
45. Graham M, Martin AMS, Peter D, James P, John WM (2008) WO2008025076
46. Masini G, Sahni S, Capellini G et al (2008) Ge photo-detectors integrated in CMOS photonic circuits. Photonics silicon III, San Jose, USA
47. Kang YM, Liu HD, Morse M et al (2009) Monolithic germanium/silicon avalanche photodiodes with 340 GHz gain-bandwidth product. Nat Photonics 3:59–63
48. Dehlinger G, Sharee J, McNab, Vlasov YA, Xia F (2009) US7515793
49. Lawrence CG, Thierry JP, Rattier MJ, Capellini G (2009) US7616904
50. Michael JH, Baehr-Jones T, Scherer A (2009) US20090052830

51. Yamada K, Tsuchizawa T, Watanabe T et al (2003) Silicon photonics based on photonic wire waveguides. Opt Letts 28:1663–1664
52. Fukazawa T, Fumiaki OHNO, Toshihiko BABA (2004) Very compact arrayed-waveguide-grating de-multiplexer using Si photonic wire waveguides. App Phys 43:673–675
53. Yamada K, Tsuchizawa T, Watanabe T et al (2009) Silicon photonics based on photonic wire waveguides. OECC 10:1–2
54. Khurgin JB (2005) Optical buffers based on slow light in electromagnetically induced transparent media and coupled resonator structures: comparative analysis. JOSAB 22:1062–1074
55. Poon JK, Zhu L, DeRose GA, Yariv A (2006) Transmission and group delay of microring coupled-resonator optical waveguides. Opt Letts 31:456–458
56. Luo X, Poon AW (2008) Double-notch-shaped microdisk resonator devices with gapless coupling on silicon chip. Chin Opt letts 7:296–298
57. Yamada K, Tsuchizawa T, Watanabe T, Fukuda H, Shinojima M, Itabashi SI (2007) Applications of low-loss silicon photonic wire waveguides with carrier injection structures. In: 4th international conference on group IV photonics, Tokyo, Japan
58. Poon AW, Xu F (2008) Silicon cross-connect filters using microring resonator coupled multi-mode-interference-based waveguide crossings. Opt Express 16:8649–8657
59. David P, Bipin D, Kalpendu S, John F, Harvey W, Margret G (2006) US20060126993
60. Ito C, Monfils I, Cartledge J, Kingston Ont (2006) All-optical 3R regeneration using higher-order four-wave mixing in a highly nonlinear fiber with a clock-modulated optical pump signal. LEOS 223–224
61. Boyraz O (2008) Nanoscale signal regeneration. Nat Photonics 2:12–13

Chapter 8
Four Wave Mixing in Silicon Photonics

Abstract In this chapter, Four Wave Mixing (FWM) and its types are elaborated in detail. It also describes the mathematical equations which provide basis for mathematical modeling and subsequent realization on various platforms. In order to achieve optical frequency shifting by FWM in silicon-on-insulator (SOI) waveguide, a nonlinear phenomenon has been presented. The FWM process in SOI waveguides is also discussed with an emphasis on the effects of two-photon absorption and the consequent free-carrier effects.

Keywords Four wave mixing • Phase matching for four wave mixing

Abbreviations

FWM	Four wave mixing
WDM	Wavelength division multiplexing
CW	Continuous wave
SPM	Self phase modulation
XPM	Cross phase modulation

List of Symbols

c	Speed of light
Tb/s	Terabits per second
Gb/s	Giga bits per second
n	Refractive Index
P	Polarization
E	Electric field
ε_0	Vacuum permittivity

J. Ahmed et al., *Optical Signal Processing by Silicon Photonics*, SpringerBriefs in Materials, 101
DOI: 10.1007/978-981-4560-11-5_8, © The Author(s) 2013

P_T	Transmitted power
α	Attenuation
dB	Decibel
χ	Tensor
P_0	Output power
μm	Micro meter
Δ	Index contrast
$1D$	One dimentional
$2D$	Two dimentional
$3D$	Three dimentional
η	Efficiency
A_{eff}	Effective area
λ	Wavelength
ω	Frequency
nm	Nano meter
P_{NL}	Nonlinear polarization
Ω_s	Frequency shift
K	Propagation constant

8.1 Four-Wave Mixing

In optical fibers, the scattering processes depend on molecular vibrations or density variations of silica. In a separate class of nonlinear phenomena, optical fibers play a passive role except for mediating interaction among several optical waves. Such nonlinear processes are referred to as parametric processes because they involve modulation of medium parameters, such as the refractive index, and require phase matching before they can build up along the fiber. Among these, FWM plays the dominant role. Although four-wave mixing (FWM) can be detrimental for wavelength division multiplexing (WDM) systems that must be designed to reduce its impact, it is also useful for a variety of applications e.g., designing light wave systems, generating a spectrally inverted signal through the process of optical phase conjugation, wavelength conversion. FWM can also be applied for phase conjugation, holographic imaging, and optical image processing.

8.2 Basis of FWM

The source of FWM lies in the nonlinear response of bound electrons of a material to an electromagnetic field. The polarization induced in the medium contains terms whose magnitude is governed by the nonlinear susceptibilities [1–3]. The resulting nonlinear effects can be classified as second- or third-order parametric processes, depending on whether the second-order susceptibility $x(2)$, or the third-order susceptibility $x(3)$, is responsible for them.

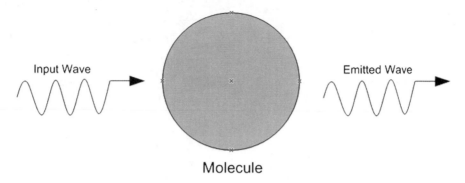

If irradiance is high enough, vibrations at all frequencies corresponding to all energy differences between populated states, are produced (Fig. 8.1).

Because $\chi^{(2)}$ vanishes for an isotropic medium (in the dipole approximation), the second-order processes such as second-harmonic generation should not occur in silica fibers. Normally they do occur because of quadrupole and magnetic-dipole effects but with relatively low conversion efficiency.

The third-order parametric processes involve nonlinear interaction among four optical waves and include the phenomena such as FWM and third-harmonic generation [1–3]. The main features of FWM can be understood from the third-order polarization term in the equation given below.

$$P_{NL} = \varepsilon_0 \chi^{(3)} : EEE$$

where E is the electric field and P_{NL} is the induced nonlinear polarization.

In general, FWM is polarization-dependent and considerable physical insight is gained by first considering the scalar case in which all four fields are linearly polarized along a principal axis of a birefringent fiber such that they maintain their state of polarization.

Fig. 8.1 Schematic diagram showing basis of four wave mixing [3]

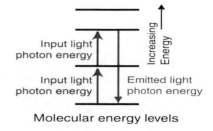

Consider four continuous waves (CWs), oscillating at frequencies $\omega_1, \omega_2, \omega_3$ and ω_4 and linearly polarized along the same axis x. The total electric field can be written as:

$$E = \frac{1}{2}x' \sum_{j=1}^{4} E_j \exp\left[i\left(k_j z - \omega_j t\right)\right] + cc$$

where the propagation constant $k_j = n_j \omega_j / c$, n_j being the mode index. Substituting the value of E from above equation in P_{NL},

$$P_{NL} = \frac{1}{2}x' \sum_{j=1}^{4} P_j \exp\left[i\left(k_j z - \omega_j t\right)\right] + cc$$

It is found that $P_j = (j = 1 \sim 4)$ consists of a large number of terms involving the products of three electric fields. For example, P_4 can be expressed as

$$P_4 = \frac{3\varepsilon_0}{4}x_{xxxx}^{(3)}\left[|E_4|^2 E_4 + 2\left(|E_1|^2 + |E_2|^2 + |E_3|^2\right)E_4\right.$$
$$\left. +2E_1 E_2 E_3 \exp(i\theta_+) + 2E_1 E_2 E_3^* \exp(i\theta_-) + \cdots\right]$$

where θ_+ and θ_- are defined as;

$$\theta_+ = (k_1 + k_2 + k_3 - k_4)z - (\omega_1 + \omega_2 + \omega_3 - \omega_4)t,$$
$$\theta_- = (k_1 + k_2 - k_3 - k_4)z - (\omega_1 + \omega_2 - \omega_3 - \omega_4)t.$$

8.3 Phase Matching Condition for FWM

The first four terms containing E_4 in P_4 are responsible for the self-phase modulation (SPM) and cross phase modulation (XPM) effects but the remaining terms result from the frequency combinations (sum or difference) of all four waves. How many of these are effective during a FWM process depends on the phase mismatch between E_4 and P_4 governed by θ_+, θ_- or a similar quantity.

If three optical fields with carrier frequencies ω_1, ω_2, and ω_3 co-propagate inside the fiber simultaneously, $\chi^{(3)}$ generates a fourth field whose frequency ω_4 is related to other frequencies by a relation $\omega_4 = \omega_1 \pm \omega_2 \pm \omega_3$. Many frequencies corresponding to different plus and minus sign combinations are possible in principle. In practice, some of these combinations do not build up because of phase matching requirement [4].

Significant FWM occurs only if the phase mismatch nearly vanishes. This requires matching of the frequencies as well as of the wave vectors. The latter requirement is often referred to as phase matching. In quantum–mechanical terms, FWM occurs when photons from one or more waves are annihilated and new photons are created at different frequencies such that the net energy and momentum are conserved during the parametric interaction. The main difference between a FWM process and a stimulated scattering process is that the phase matching condition is automatically satisfied in the case of Raman or Brillouin scattering as a result of the active participation of the nonlinear medium. In contrast, the phase matching condition requires a specific choice of input wavelengths and fiber parameters before FWM can occur with high efficiency.

The phase matching condition can be approximately satisfied if the zero-dispersion wavelength of the fiber is chosen to coincide with the pump wavelength. The fiber nonlinearity generates the phase-conjugated signal at the frequency $\omega_c = 2\omega_p - \omega_s$, provided that the phase matching condition $k_c = 2k_p - k_s$.

8.4 Degenerate and Nondegenerate FWM

There are two types of FWM terms in P_4, the term containing θ_+ corresponds to the case in which three photons transfer their energy to a single photon at the frequency $\omega_4 = \omega_1 + \omega_2 + \omega_3$. This term is responsible for the phenomena such as third-harmonic generation ($\omega_1 = \omega_2 = \omega_3$), or frequency conversion when $\omega_1 = \omega_2 \neq \omega_3$. In general, it is difficult to satisfy the phase matching condition for such processes to occur in optical fibers with high efficiencies. FWM is also present if only three waves interact. In this case the term $F0 = f1 + f1 - f2$ couples three components, thus generating the so-called Degenerate FWM, showing identical properties as in case of four interacting waves.

FWM can take place in any material; it refers to interactions of the waves via a third-order nonlinear polarization. Degenerate four waves mixing (DFWM) can yield phase conjugation and is useful, for example, for correcting aberrations by using a phase conjugate mirror (PCM).

FWM is a fiber-optic characteristic that affects WDM systems, where multiple optical wavelengths are spaced at equal intervals or channel spacing. The effects of FWM are pronounced with decreased channel spacing of wavelengths and at high-signal power levels. High chromatic dispersion decreases FWM effects, as the signals lose coherence. The interference FWM causes in WDM systems is known as interchannel crosstalk. FWM can be mitigated by using uneven channel spacing or fiber that increases dispersion.

The term containing θ_- corresponds to the case in which two photons at frequencies ω_1 and ω_2 are annihilated, while two photons at frequencies ω_3 and ω_4 are created simultaneously such that:

$$\omega_3 + \omega_4 = \omega_1 + \omega_2$$

The phase matching requirement for this process is $\Delta k = 0$, where

$$\Delta k = k_3 + k_4 - k_1 - k_2$$
$$= (n_3\omega_3 + n_4\omega_4 - n_1\omega_1 - n_2\omega_2)/c = 0$$

and n_j is the effective mode index at the frequency ω_j (Fig. 8.2).

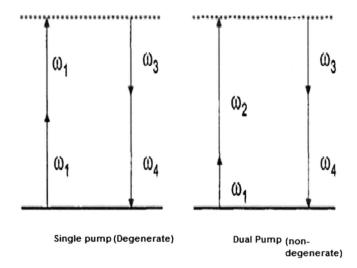

Single pump (Degenerate)   Dual Pump (non-
                                                              degenerate)

Fig. 8.2 Energy levels of degenerate and nondegenerate FWM

Fig. 8.3 Generation of new
optical waves

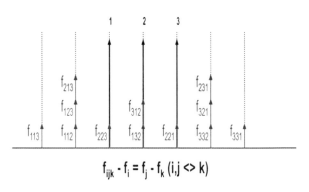

$$f_{ijk} - f_i = f_j - f_k \ (i,j <> k)$$

In the general case where $\omega_1 \neq \omega_2$, one must launch two pump beams for FWM to occur. The special case, in which $\omega_1 = \omega_2$ is interesting because FWM can be initiated with a single pump beam. This degenerate case is often useful for optical fibers. Physically, it manifests in a way similar to SRS. A strong pump wave at ω_1 creates two sidebands located symmetrically at frequencies ω_3 and ω_4 with a frequency shift

$$\Omega_s = \omega_1 - \omega_3 = \omega_4 - \omega_1$$

Where it is assumed for definiteness that $\omega_3 < \omega_4$. The low-frequency sideband at ω_3 and the high-frequency sideband at ω_4 are referred to as the Stokes and anti-Stokes bands in direct analogy with SRS. The degenerate FWM was originally called three wave mixing as only three distinct frequencies are involved in the nonlinear process. The Stokes and anti-Stokes bands are often called the signal and idler waves.

Explaining the Fig. 8.3 in which due to interaction of the transmitted optical waves, the mixing products interfere with the transmitted channels causing consequent eye closing and BER degradation.

The important issue in design of WDM light wave systems is the interchannel crosstalk. The system performance degrades whenever crosstalk leads to transfer of power from one channel to another, such a transfer can occur because of non-linear effects in optical fibers. The XPM and SPM both affect the performance of WDM systems.

8.5   Wavelength Conversion by FWM in SOI Waveguides

8.5.1   Wavelength Conversion Schemes

A high-speed wavelength converter is an essential part within a high capacity all optical WDM system and thus considerable interest in the development of a practical wavelength converter exists. All optical wavelength converter satisfies the high-speed requirement of modern communication systems and can be used for single channel wavelength conversion and wavelength conversion of a WDM signal (Figs. 8.4 and 8.5).

Reports on optical wavelength conversion using various methods already exist: XPM, super continuum generation and spectral slicing, FWM in a fiber and interferometric SOA wavelength conversion, just to name a few. An ideal optical wavelength converter is transparent to both data rate and modulation format. One method to get such an optical wavelength converter is to use FWM in a nonlinear waveguide.

Modern high-speed optical systems are likely to use an advanced modulation formats such as differential phase shift keying (DPSK) and return to zero DPSK (RZ-DPSK) due to their various advantages compared with conventional on–off

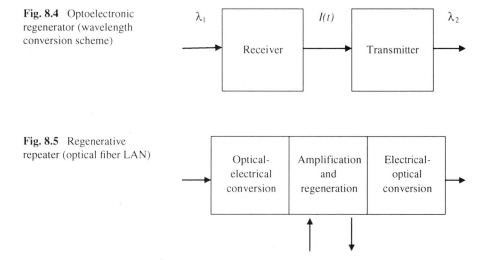

Fig. 8.4 Optoelectronic regenerator (wavelength conversion scheme)

Fig. 8.5 Regenerative repeater (optical fiber LAN)

keying (OOK) format, for example better receiver sensitivity associated with balanced detection, larger dispersion tolerance and better resilience to fiber nonlinear effects. With the recent interest in phase-modulated signals for optical communication systems, it is interesting to look at wavelength conversion methods suitable for such signals.

Parametric amplification is promising because of their large and flexible gain bandwidth. Operation of parametric amplification with CW pumps would be desirable for many practical applications. This would provide gain or idler conversion efficiency independent of time, which is convenient for amplifying or converting input signals with arbitrary modulation formats. However, recently available highly nonlinear silicon waveguides have a zero-dispersion wavelength, which can vary by several nanometers over a few centimeters. This in turn prevents the phase matching condition from being maintained at its optimum value along the waveguide. As a result, to date, CW parametric amplifiers have not exhibited gain spectra as wide as that by theory. In contrast, by using a pulsed pump, which consists of high-peak power pulses, and relatively short waveguide length, one can obtain very wide (>200 nm) gain spectra, with a shape very close to that predicted by theory.

8.5.2 Wavelength Conversion by FWM in SOI Waveguides

The flexibility of very high capacity optical fiber networks solely depends upon all optical wavelength conversion. Among many wavelength conversion schemes, exploiting FWM [5] in silicon waveguides has gained considerable attraction due to its transparency to modulation formats and bitrates, along with its high integration and compatibility with electronic devices. The silicon waveguides are the good candidate for FWM applications due to unbelievable reduction in cost and size of silicon photonic devices.

The parametric process of FWM in silicon waveguides [6] is the result of nonlinear response of bound electrons to an applied field. The inherent properties of very high nonlinearity and tight light confinement of silicon waveguides enable it to provide silicon compatible chip scale wavelength converters, ready to replace the long highly nonlinear-silica fibers (HNLSF).

Utilization of nondegenerate FWM in semiconductor optical amplifiers (SOA) is a common approach to realize frequency shifting [7–9] or dispersion shifted silica fibers may be used. Recently, wavelength conversion for 10, 20, 40, and 160 Gb/s has been demonstrated in SOI waveguides [10–12]. In addition, other signal processing devices such as optical modulation [13], optical switching [14, 15], optical amplification [16], and lasing [17] have also been presented, benefitting with the recent advancements in CMOS-compatible silicon photonics.

In this chapter, it has been investigated that the conversion efficiency in accordance with the pump/signal powers and wavelengths for 1 cm long silicon waveguide while taking into account the noise and losses as noise figure (NF).

By taking the coupled-amplitude wave equations that govern the nonlinear dynamics of two optical waves interacting inside a SOI waveguide.

Generally, these equations consider not only linear loss, dispersion, and the free-carrier effects, but also take into account for the intensity-dependent nonlinear losses due to two photon and free-carrier absorption. Using approximations based on physical insights, we simplify the equations as per situations of practical interest and outline techniques that can be used to examine the impact of nonlinear behavior as light propagates through a 1 cm silicon waveguide. In particular, propagation of single pump and a signal through a waveguide of constant cross section is described. A short description about the noise and losses are also discussed in this section and NF is estimated for silicon waveguide under subject.

8.6   Mathematical Model of Wavelength Conversion in SOI Waveguides

8.6.1   Propagation of Light in Silicon Waveguides

FWM is a process in which three or four waves co-propagate simultaneously inside a silicon waveguide. Four waves are involved in the nondegenerate FWM [18, 19], while only three waves are present in the case of degenerate FWM [20]. These waves are pump, signal, and idler waves as shown in Figure, and there photon energies satisfy the energy conservation relation (Fig. 8.6)

$$2h\omega_p = h\omega_s + h\omega_i \tag{8.1}$$

In this book, the focus is on the degenerate FWM configuration, where only a single pump wave is involved along with a signal and idler waves. These are assumed to be identically polarized in either the fundamental TE or TM mode. The TE and TM modes have a polarization component along the 'z' direction so only a small fraction of incident power is present and only the transverse component contains the nonlinear effects. During their propagation, the two optical fields induce material polarization $P(r,t)$. This polarization drives the evolution of total electric field $E(r,t)$ according to the Maxwell wave equation [4].

$$\nabla^2 E - \frac{1}{c^2}\frac{\partial^2 E}{\partial t^2} = \frac{1}{\varepsilon_0 c^2}\frac{\partial^2 P}{\partial t^2} \tag{8.2}$$

where ε_0 is the free-space permittivity and c is the speed of light in vacuum. The linear and nonlinear polarizations constitute the total polarization in silicon as

$$P(r,t) = P_L(r,t) + P_{NL}(r,t)$$

The main contribution sources to nonlinear polarization in silicon are the refractive index changes induced by the photo-generated free carriers, stimulated Raman scattering (SRS), and polarization of bound electrons [21].

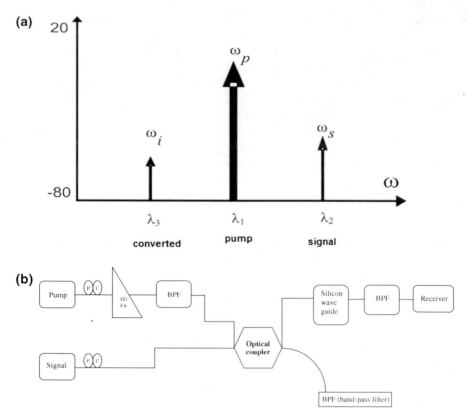

Fig. 8.6 **a** A typical wavelength conversion by FWM. **b** The schematic configuration of the wavelength converter in silicon waveguide [20]

By assuming that the reflections are very weak so that backward propagating waves can be safely discarded, it is supposed that the optical waves propagate only in the forward, $+z$, direction. The total electrical field inside silicon waveguide will be in the form

$$E(r,t) = s_p(z)F_p(r)A_p(z,t)\exp[i(\beta_{0p}z - \omega_p t)]$$
$$+ s_s(z)F_s(r)A_s(z,t)\exp[i(\beta_{0s}z - \omega_s t)] + c.c.$$

where $\omega_m = 2\pi c/\lambda_m$ ($m = p$ or s for pump and signal waves) is the carrier frequency, $\beta_{0m} = n_{0m}k_m$ is the propagation constant, $k_m = \omega_{m/c}$ is the free-space wave number, and n_{0m} is the effective refractive index. To account for the waveguide tapering, here introduced $s_m(z)$ as the shape function [22].

8.6.2 Wavelength Conversion

By means of a standard procedure [23], the wave Eq. (8.2) can be reduced to the system of the following (8.3–8.5) coupled-amplitude equations that govern the evolution of the pump and signal envelopes inside the SOI waveguide as a result of FWM parametric process. It can result in amplification of the signal as well as generation of an idler wave (ω_i) at the frequency $\omega_i = 2\omega_p - \omega_s$ [24, 25]. Assuming that $|Ap| > |As| > |Ai|$

$$\frac{dA_p}{dz} = -\frac{1}{2}\left[\alpha + \alpha_p^{FCA}(z)\right]A_p + i\left(\gamma_p + i\frac{\beta}{2}\right)|A_p|^2 A_p \qquad (8.3)$$

$$\frac{dA_s}{dz} = -\frac{1}{2}\left[\alpha + \alpha_s^{FCA}(z)\right]A_s + 2i\left(\gamma_s + i\frac{\beta}{2}\right)|A_p|^2 A_s + i\gamma_s A_p^2 A_i^* exp(-i\Delta k \cdot z)$$
$$(8.4)$$

$$\frac{dA_i^*}{dz} = -\frac{1}{2}\left[\alpha + \alpha_i^{FCA}(z)\right]A_i^* - 2i\left(\gamma_i + i\frac{\beta}{2}\right)|A_p|^2 A_i^* - i\gamma_i A_p^{2*} A_s exp(i\Delta k \cdot z) \quad (8.5)$$

The first term on right-hand side of Eqs. (8.3–8.5) is responsible for attenuations due to linear absorption and free-carrier absorption, the second term is responsible for SPM and TPA in Eq. (8.3) and XPM and TPA in Eqs. (8.4) and (8.5), and the last term in Eqs. (8.4) and (8.5) describes the energy transfer between the interacting waves. The use of linear loss coefficient $\alpha = 1.4$ dB/cm the TPA coefficient $\beta = 0.75 \times 10^{-11}$ m/W and the nonlinearity coefficient $\gamma_j = n_2\omega_p/c$ with the nonlinear refractive index $n_2 = 5.5 \times 10^{-18}$ m^2/W $(j = p,s,i)$. Δk is the phase mismatch due to propagation constants. The TPA-induced FCA loss is given as $\alpha_j^{FCA}(z) = 1.45 \times 10^{17} (\lambda_j/1,550)^2 N$, where λ_j is the wavelength (nm), $N (cm^{-1})$ is carrier density generated by TPA. Here N should satisfy the following rate equation at any position of the waveguide at any time [26].

$$\frac{dN(t,z)}{dt} = \frac{\beta}{2hv}I^2(t,z) - \frac{N(t,z)}{\tau} \qquad (8.6)$$

In this equation, I is the peak intensity, hv is the photon energy, and τ is the effective carrier lifetime, which changes with the waveguide geometry or reverse bias voltage if a p-i-n diode structure exists. For CW pumping or long pulse pumping, N will reach the local steady-state value of $N(z) = \tau\beta I^2(z)/hv$. For pulse pumping, the repetition rate R of the pulsed pump is an important factor impacting N. Under the operating condition of pulse pumping with pulse width $T_0 << \tau$, N is given by

$$N(t,z) \approx \frac{\left(\frac{1}{1-e^{-1/R\tau}}\right)\beta T_0 I^2(t,z)}{2hv} \qquad (8.7)$$

FWM is a coherent process whose efficiency depends on how well the phase mismatch meets the phase matching condition

$$\begin{aligned}\Delta k &= k_s + k_i + 2k_p + 2\gamma P_p \\ &= \left(n_s\omega_s + n_i\omega_i - 2n_p\omega_p\right)/c + 2\gamma P_p \\ &= \beta_2\left(\omega_s - \omega_p\right)^2 + 2\gamma P_p \\ &= 0\end{aligned} \qquad (8.8)$$

where k_j is the propagation constant and β_2 is the group velocity parameter. As the nonlinear part of the phase match is positive, phase matching can be realized by locating the pump wave in the anomalous dispersion regime ($\beta_2 < 0$) so that the linear phase mismatch can compensate the nonlinear one. At the telecommunication wavelength of 1,550 nm, the material dispersion of crystalline silicon is normal ($D = -2\pi c\beta_2/\lambda^2 = -880\, ps/(\text{nm} \times \text{km})$) Because of the strong modal confinement in silicon waveguides, the waveguide dispersion can counteract the effect of the normal material dispersion [27]. It is verified simulationsly that through proper design of waveguide shape and size, anomalous GVD in the range of $200 - 1,200\, ps/\text{nm} \times \text{km}$ can be obtained at 1,550 nm [28].

In addition to waveguide geometry and material dispersion, the TPA-induced free carrier dispersion (FCD) may also influence the phase matching condition by changing the refractive index locally. Due to the free-carrier plasma effect, the refractive index of silicon decreases linearly with increasing carrier density. However, the quadratic wavelength dependence of the free-carrier plasma effect

$$\Delta n_{FC} = -\frac{e^2\lambda^2}{8\pi^2 c^2 \varepsilon_0 n}\left(\frac{\Delta N_e}{m^*_{ce}} + \frac{\Delta N_h}{m^*_{ch}}\right) \approx -8.2 \times 10^{-22}\lambda^2 N \qquad (8.9)$$

will alter the local dispersion and phase mismatching conditions as described in the following FCD equation:

$$\Delta D_{FC} = \frac{1}{c}\frac{d\Delta n_{FC}}{d\lambda} = -5.46 \times 10^{-30}\lambda N \qquad (8.10)$$

8.7 Noise Figure

In silicon waveguides, the gain is generated by FWM process and the attenuation is caused by nonlinear losses of TPA, FCA, and linear loss. The noise is associated with photon fluctuation created by the gain and the loss in the optical amplification process. Noise is a term generally used to refer to any undesired disturbances that mask the received signal in a communication system. The NF is thus the ratio of actual output noise to that which would remain if the device itself did not introduce any type of noise. It is a number by which the performance of a radio receiver can be specified.

There are three main types of noise due to spontaneous fluctuations in optical fiber communication system;

- Thermal noise
- Dark current noise
- Quantum noise.

The shot noise and thermal noise are the two fundamental noise mechanisms responsible for current fluctuations in all optical receivers even when the incident optical power P_{in} is constant.

One can find detailed derivation for noise calculations in silicon waveguides in, but here noise induced by FWM process and contributions from other components are considered to estimate the NF which impacts our simulations. To calculate the NF due to FWM process, mean output photon number and mean photon number fluctuations through the silicon waveguide are calculated using the signal wave Eqs. (8.3–8.5), and amplification based on photon fluctuations will be:

$$NF_{silicon} = \frac{T + N_{loss} + N_{gain}}{T} + \frac{N_{gain}(T + N_{loss})}{T^2 |a|^2} \tag{8.11}$$

where

$$T = exp\left(\int_0^L (g(z) - l(z) dz)\right) \tag{8.11a}$$

is the net gain and

$$N_{gain} = \int_0^L g(z) exp\left(\int_z^L (g(x) - l(x) dx)\right) dz \tag{8.11b}$$

$$N_{loss} = \int_0^L l(z) exp\left(\int_z^L (g(x) - l(x) dx)\right) dz \tag{8.11c}$$

are the photon fluctuations due to gain and loss. Parameter L is the waveguide length, $g(z)$ is the gain parameter that can be numerically calculated from Eqs. (8.3–8.5). $l(z) = \alpha + \alpha^{FCA}(z) + 2\beta l(z)$ is the experienced loss coefficient and $|a|^2$ is the photon number at the input frequency. Considering high-input signal power lets the second term in Eq. (8.11) negligible and (8.11a–8.11c) can be numerically solved. Neglecting the pump relative intensity noise (RIN) the total NF will be

$$NF = NF_{silicon} + NF_p \tag{8.12}$$

In a 1 cm silicon waveguide operating in anomalous dispersion regime at 1,550 nm, the typical NF spectra of the pump with 1-ps pulses operating at repetition rate of 10 GHz are given in Fig. (8.7). as illustrated, the NF contribution from gain and loss fluctuations in the silicon waveguide is the dominant noise source of the silicon FWM process.

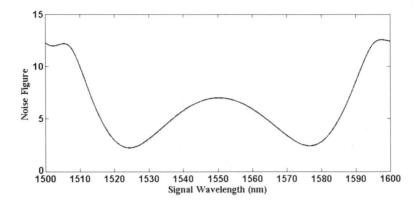

Fig. 8.7 NF spectra in SOI against the signal wavelength for 1 cm long waveguide

References

1. Schubert M, Wilhelmi B (1986) Nonlinear optics and quantum electronics. Wiley, New York
2. Shen YR (1984) The principles of nonlinear optics. Wiley, New York
3. Boyd RW (2003) Nonlinear optics, 2nd edn. Academic Press, San Diego
4. Agrawal GP (2001) Nonlinear fiber optics, 3rd edn. Academic Press, San Diego
5. Raghunathan V, Claps R, Dimitropoulos D, Jalali B (2004) Wavelength conversion in silicon using Raman induced four-wave mixing. Appl Phys Letts 85:26–34
6. Agrawal GP (2007) Nonlinear fiber optics, 4th edn. Academic Press, Boston
7. Tiemeijer LF (1991) Effects of nonlinear gain on four-wave mixing and asymmetric gain saturation in a semiconductor laser amplifier. Appl Phys Lett 59:499–501
8. Tatham MC, Sherlock G, Westbrook LD (1993) 20-nm optical wavelength conversion using nondegenerate four-wave mixing. IEEE Photon Technol Lett 5:1303–1306
9. Murata S, Tomita A, Shimizu J, Suzuki A (1991) "THz optical-frequency conversion of 1 Gb/s signals using highly nondegenerate four-wave mixing in an InGaAsP semiconductor laser". IEEE Photon Technol Lett 3:1021–1023
10. Nunes LR, Liang TK, Tsuchiya M, Van Thourhout D, Dumon P, Baets R (2005) Ultrafast noninverting wavelength conversion by crossabsorption modulation in silicon wire waveguides. In: Proceedings of 2nd IEEE international conference group IV photon, pp 154–156
11. Foster MA, Turner AC, Lipson M, Gaeta AL (2008) Nonlinear optics in photonic nanowires. Opt Exp 16:1300–1320
12. Tsang HK, Liu Y (2008) Nonlinear optical properties of silicon waveguides. Semicond Sci Technol 23:064007-1–064007-9
13. Preble SF, Xu Q, Schmidt BS, Lipson M (2005) Ultrafast all-optical modulation on a silicon chip. Opt Lett 30:2891–2893
14. Almeida VR, Barrios CA, Panepucci RR, Lipson M, Foster MA, Ouzounov DG, Gaeta AL (2004) All-optical switching on a silicon chip. Opt Lett 29:2867–2869
15. Vlasov Y, Green WMJ, Xia F (2008) High-throughput silicon nanophotonicwavelength-insensitive switch for on-chip optical networks. Nat Photon 2:242–246
16. Foster MA, Turner AC, Sharping JE, Schmidt BS, Lipson M, Gaeta AL (2006) Broad-band optical parametric gain on a silicon photonic chip. Nature 44:960–963
17. Boyraz O, Jalali B (2004) Demonstration of a silicon Raman laser. Opt Exp 12:5269–5273
18. Tatham MC, Sherlock G, Westbrook LD (1993) 20 nm optical wavelength conversion using nondegenerate four-wave mixing. IEEE Photon Technol Lett 5:1303–1306

19. Zhou J, Park N, Vahala KJ, Newkirk MA, Miller BI (1994) Study of Interwell carrier transport by terahertz four-wave mixing in an optical amplifier with tensile and compressively strained quantum wells. Appl Phys Lett 65:1897–1899
20. Zhang J, Lin Q, Agrawal GP, Fauchet PM (2006) Broadband optical amplification and wavelength conversion by four-wave mixing in silicon waveguides. In: 3rd IEEE Group IV photonics
21. Lin Q, Painter OJ, Agrawal GP (2007) Nonlinear optical phenomena in silicon waveguides: modelling and applications. Opt Exp 15:16604–16644
22. Rukhlenko ID, Premaratne M, Agrawal GP (2010) Nonlinear silicon photonics: analytical tools. IEEE J Sel Top Quantum Electron 16(1):200–215
23. Chen X, Panoiu NC, Osgood RM (2006) Theory of Raman-mediated pulsed amplification in silicon-wire waveguides. IEEE J Quantum Electron 42(2):160–170
24. Soref RA, Lorenzo JP (1986) All-silicon active and passive guided-wave components for $\lambda = 1:3$ and $1:6$ μm. IEEE J Quantum Electron 22:873–879
25. Pavesi L, Lockwood D (2004) Silicon photonics: topics in applied physics, vol 94. Springer, Berlin
26. Boyraz O, Koonath P, Raghunathan V, Jalali B (2004) All optical switching and continuum generation in silicon waveguides. Opt Express 12:4094–4102
27. Sang X, Boyraz O (2008) Gain and noise characteristics of high-bit-rate silicon parametric amplifiers. Opt Express 16:13122–13132
28. Turner AC, Manolatou C, Schmidt BS, Lipson M, Foster MA, Sharping JE, Gaeta AL (2006) Tailored anomalous group-velocity dispersion in silicon channel waveguides. Opt Express 14:4357–4362

Chapter 9
FWM Implementation and Analysis Using MATLAB

Abstract The third-order nonlinear process of four-wave mixing (FWM) in silicon-on-insulator (SOI) waveguides has been investigated intensively due to the large third-order susceptibility of silicon. The exploration of nonlinear optical processes in SOI waveguides has achieved a level from where the researchers can move ahead of the proof-of-concept stage and enter the practical application empire. Signal amplification, wavelength conversion, and signal regeneration are exceptionally available applications in the SOI-platform for present and future integrated photonic circuits. These applications are only possible due to the large pump powers and small modal areas of the SOI waveguides which are equally essential for large nonlinear optical responses. Further investigations can produce silicon photonic devices for optical signal processing functions, i.e., modulation, logic operations, optical signal regeneration, spectral phase conjugation, signal-waveform compression, and a number of other areas. In this chapter, the utility of FWM in SOI waveguides for all-optical wavelength conversion is investigated by exploiting the Simulink MATLAB tool. Realization of signal amplification with wavelength conversion is demonstrated and then optimized by analyzing influence of various parameters of pump, signal, and waveguide. For the majority of applications of FWM in SOI waveguides, the most required phenomenon of phase matching has been taken into consideration during each simulation step.

Keywords Phase matching • All-optical signal processing • Wavelength conversion • Phase matching • Silicon photonic devices • Repetition rate

Abbreviations

FWM   Four wave mixing
SOI   Silicon-on-insulator
CE   Conversion efficiency
TPA   Two photon absorption

J. Ahmed et al., *Optical Signal Processing by Silicon Photonics*, SpringerBriefs in Materials,   117
DOI: 10.1007/978-981-4560-11-5_9, © The Author(s) 2013

XPM   Cross phase modulation
FCA   Free carrier absorption
GVD   Group velocity dispersion
NF   Noise figure

9.1 Implementation Methodology

Continuation to Chaps. 7 and 8 where mathematical background and procedure for implementation of FWM in SOI waveguides are discussed in detail, following are the tuning parameters considered during simulations.

Pump Parameters

- Pump Power
- Pump wavelength
- Pump-pulsed operation
- Pump pulse width
- Angular frequency
- Pump initial power
- Pump polarization
- Loss due to pump.

Signal Parameters

- Signal wavelength
- Signal bit rate
- Signal pulse width
- Signal bandwidth
- Angular frequency
- Signal initial power
- Loss due to signal
- Signal polarization.

SOI Waveguide Parameters

- Waveguide structure
- Waveguide Dispersion
- Waveguide effective area

- Waveguide length
- Waveguide linear refractive index
- Waveguide nonlinear refractive index.

Other Tuning Parameters

- Linear loss coefficient
- Nonlinear loss coefficient
- Number of free carriers
- Two photon absorption coefficient
- Free carriers absorption coefficient
- Stokes power
- Conjugate power
- Phase mismatch coefficient
- Linear noise
- Nonlinear noise
- Total noise
- Energy of photon
- Net gain
- Net loss
- Noise figure.

Considering the above-mentioned key parameters, the investigated research-results obtained by using simulation are discussed and demonstrated below. Mostly the pump power in presence of different free carriers' life time is simulated. The dispersion values along with wavelengths and wavelength detuning (between pump and signals) impact on conversion efficiency is investigated. The impact of waveguide length, pump pulse-width, and repetition rate are presented in particular in this section.

To analyze the performance of frequency, the impact of operating wavelengths and pump pulse parameters are investigated numerically. The silicon waveguide under discussion has been considered to be 1 cm long with $0.1\ \mu m^2$ effective area.

9.2 Impact of Pump Power

The different pump powers influence the phase matching conditions according to Eqs. (8.8) and (8.10). Figures 9.1 and 9.2 show the influence of the pump power on the CE. It is obvious that the CE increases with the increase of pump power first and then gradually saturates or even decreases with the further increase in the pump power. This is due to the effective carrier lifetime and to the losses, especially the FCA loss, which causes the effective pump power to reduce the FWM process.

It can be seen in Figs. 9.1 and 9.2 that the CE decreases with an increase in the effective carrier lifetime, which mainly increases the accumulation of the carriers and therefore increases the FCA loss. Even with ($\tau = 0$ ps, red curve) the plot without considering the effect of the FCA also shows the saturation against increasing pump power. This is attributed to the TPA which becomes evident

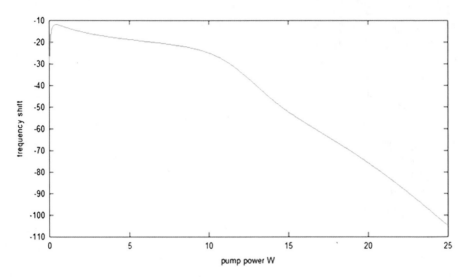

Fig. 9.1 Frequency shift vs. signal power

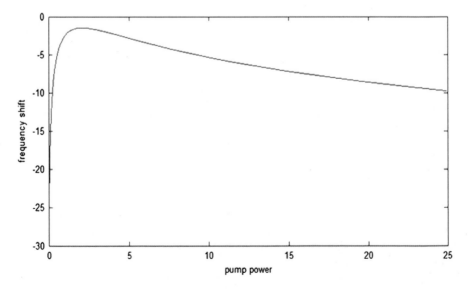

Fig. 9.2 Effect of dispersion

when the pump power is high. So increasing the pump power does not always result in the increase of the CE, which is not only dependent on the range of the pump power but also on the lifetime of the carriers. As it is well identified that with increasing pump power the NF starts increasing but this increase is different for different wavelengths, e.g., the NF degrades rapidly for wavelengths near 1,550 nm and slowly for wavelengths near 1,510 nm. This ultimately shows the dependency of CE on wavelength at higher pump powers. Hence the optimum values of pump power, pump/signal wavelength, and the waveguide length are essential to achieve improved NF and maximum CE.

In this context, the pump power plays a significant role in wavelength conversion efficiency in silicon waveguides, also TPA and FCA are both intensity-dependent phenomenon.

9.3 Impact of Wavelength

As in the previous work, due to nonlinear losses and phase mismatching, the wavelengths near 1,500 nm achieve higher gain and hence efficiency as compared to those near 1,550 nm at peak powers. The wavelength detuning is defined in this work as the wavelength difference of the signal wavelength to the pump wavelength $\Delta\lambda = \lambda_p - \lambda_s$. The effect of wavelength detuning on CE is simulated and found that the results for positive and negative $|\Delta\lambda|$ are almost the same. As shown in Fig. (9.3) the CE is optimum at $\Delta\lambda \approx 20\,\mathrm{nm}$ and it lowers for both the higher and lower values of detuning. The drop of CE at lower detuning values is due to the phase mismatch and nonlinear losses while at higher detuning values the drop in CE is caused by the double peaks observed in Fig. 9.3. Figure 9.4 concludes the

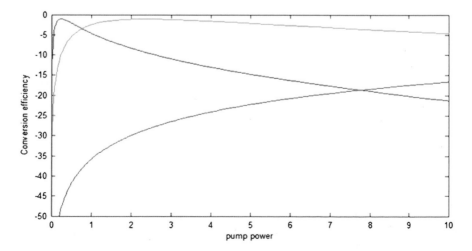

Fig. 9.3 Effect of effective area

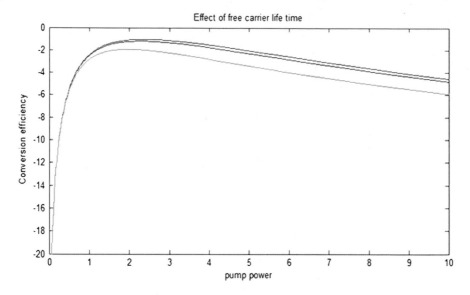

Fig. 9.4 Effect of free carrier life time

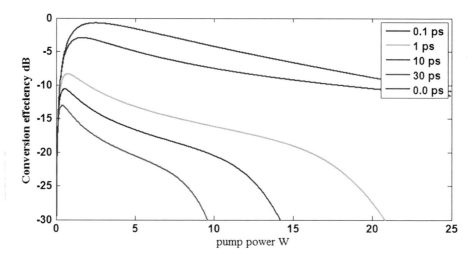

Fig. 9.5 Effect of free carrier life time and conversion efficiency

impact of wavelength-dependent NF on CE. It is obvious that the pump and signal wavelengths should be chosen carefully along with other parameters to attain phase matching conditions required for FWM and ultimately the optimum wavelength conversion efficiency (Figs. 9.5, 9.6 and 9.7).

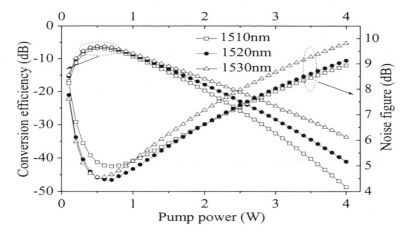

Fig. 9.6 Conversion efficiency profiles at different wavelengths against pump power

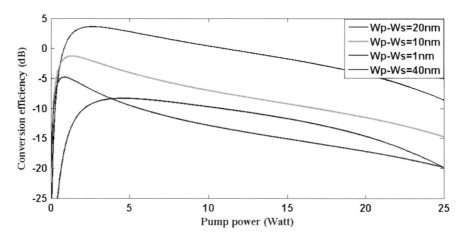

Fig. 9.7 Conversion efficiency evolution at different wavelength detuning values (W_p-W_s), W_p is pump wavelength and W_s is signal wavelength

9.4 Waveguide Length, Pulse Width, and Repetition Rate

The pump pulse-width and repetition rate showed noticeable impact on free carrier density and hence the conversion efficiency. Higher the repetition rate and wider the pump pulses results in lowering the CE. The propagation of the pulse in the waveguide is governed by the interplay of the linear dispersion and nonlinearity. These effects can be described in terms of waveguide characteristic lengths namely the GVD length, the TOD length, and the nonlinear length. The nonlinear length of a silicon waveguide is seen to be several orders of magnitude shorter than optical fibers, for a typical peak power. The short nonlinear length is due to very high nonlinear parameter, also dispersion and nonlinear lengths are pulse-width dependent. Thus, the dispersion length for a usual ps pulse is several orders of magnitude longer than that for an fs pulse, but the nonlinear length is the same for both short and long pulses as long as the pulse power is the same (Fig. 9.8).

The impact of waveguide length is shown in Fig. 9.9, while the combined resulted curves for CE and NF are depicted in Fig. 9.9. The free carrier time used in simulation is 1,000 ps for same waveguide length of 1 cm.

The curves clearly show that the increase in NF decreases the CE. The conversion efficiency for longer waveguide lengths is lower at the same pump powers while for short waveguides, it is higher. As in above figure the CE for 1, 1.5, and 3 cm is −2, −4, and −8 dB, respectively.

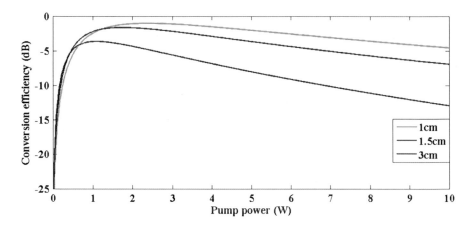

Fig. 9.8 Conversion efficiency at different waveguide lengths

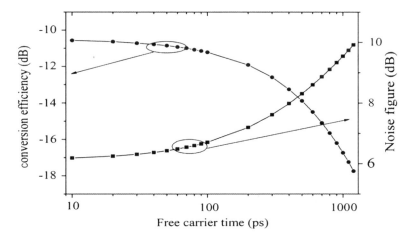

Fig. 9.9 Noise figure increases with the increase in power and waveguide length, which strongly degrades the conversion efficiency

9.5 Comments on Obtained Results and Future Recommendations

The influence of pump power, dispersion, wavelength, and nonlinear absorption on the conversion efficiency is discussed in presence of noise figure as depicted in Fig. 9.9. In order to achieve high conversion efficiency, the pump wavelength should be operated in anomalous GVD regimes. Conversion efficiency in silicon waveguides decreases with the increasing pump power and the noise figure is degraded due to the two photon absorption (TPA) and TPA-induced FCA at the high pump power. With the increase in free carrier lifetime, conversion efficiency will decrease and the noise figure will increase accordingly. In practical application, the high conversion efficiency and low noise figure can be achieved by choosing suitable parameters. It is evident that the nonlinear loss phenomenon of TPA and FCA are dominant at low pump powers, with increased pump power the FWM becomes phase matched at the wavelength under consideration. For wavelengths 20–30 nm far from the pump, the gain overcomes nonlinear losses resulting in higher CE. Here the noise figure is strongly associated with carrier recombination life time and the life time should be of the order of 300 ps or less (Fig. 9.10).

Several kinds of nonlinear optical effects have been observed in silicon waveguides in recent years, and their device applications are attracting considerably attention. In this work, it is provided a unified platform that can be used for understanding the underlying physics and guidance toward new and useful applications. Begin with a description of SOI waveguides and the third-order nonlinearity of silicon. The generation of free carriers through TPA and their impact on various nonlinear phenomena is included fully in the simulation work presented here.

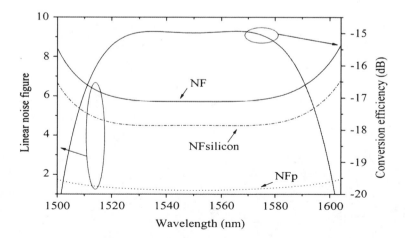

Fig. 9.10 Conversion efficiency and *NF* spectra of wavelength converter in silicon waveguide

By using the coupled amplitude equations to study interaction of optical pulses in the SOI waveguides and showed the optical wavelength conversion. The non linear phenomenon of FWM is used to generate wavelength conversion. It is considered first that the impact of free carriers and showed that, index changes induced by them have a significant impact on FWM.

The influence of pump power, dispersion, wavelength, and nonlinear absorption on the conversion efficiency is discussed in presence of noise figure. In order to achieve high conversion efficiency, the pump wavelength should be operated in anomalous GVD regimes. Conversion efficiency in silicon waveguides decreases with the increasing pump power and the noise figure is degraded due to the TPA and TPA-induced FCA at the high pump power. With the increase in free carrier lifetime, conversion efficiency will decrease and the noise figure will increase accordingly. In practical application, the high conversion efficiency and low noise figure can be achieved by choosing suitable parameters. It is evident that the non-linear loss phenomenon of TPA and FCA are dominant at low pump powers, with increased pump power the FWM becomes phase matched at the wavelength under consideration. For the wavelengths 20–30 nm far from the pump, the gain over-comes nonlinear losses resulting in higher CE. Noise figure is strongly associated with carrier recombination life time and the life time should be of the order of 300 ps or less. Given simulation shows the Bit rate around 10 GHZ, pulse width 1 ps, signal wavelength 1,530 nm for a SOI waveguide length of 1 cm are better choice for optimum conversion efficiency.

The simulation shows the Bit rate around 10 GHZ, pulse width 1 ps, signal wavelength 1,530 nm for a SOI waveguide length of 1 cm are better choice for optimum conversion efficiency. The further improvement can be achieved if TPA-induced FCA loss is reduced and NF improved in silicon waveguides.

Printed by Publishers' Graphics LLC
DBT131107.15.14.199